DEPARTMENT OF THE INTERIOR

ALBERT B. FALL, Secretary

UNITED STATES GEOLOGICAL SURVEY

GEORGE OTIS SMITH, Director

Bulletin 732

GEOLOGY AND ORE DEPOSITS OF SHOSHONE COUNTY, IDAHO

BY

JOSEPH B. UMPLEBY

AND

E. L. JONES, JR.

WASHINGTON

GOVERNMENT PRINTING OFFICE

1923

II

CONTENTS.

ILLUSTRATIONS.

GEOLOGY AND ORE DEPOSITS OF SHOSHONE COUNTY, IDAHO.

By Joseph B. Umpleby and E. L. Jones, Jr.

INTRODUCTION.

The Coeur d'Alene district, the most productive mining district in Idaho and one of the great lead-silver districts of the world, has undergone rapid development during the 15 years since it was studied by Ransome and Calkins.[1] This fact, together with added information concerning the geology and mineral deposits in adjacent areas, is the justification for this report on the ore deposits of Shoshone County. It was hoped that by studying the deeper ore bodies opened since the previous examination and by viewing the deposits in a broader setting of regional information new light might be thrown upon their origin and a better basis offered for intelligent exploration in adjacent areas.

The investigation was originally planned as part of a broad study of the ore deposits of Idaho, upon which the senior author had spent several field seasons. Owing to interruption of the work during the World War and the subsequent resignation of both authors from the Survey the original plan was abandoned. The investigation of Shoshone County, however, represented the only important part of the broad study that had not been recorded in official publications, so the senior author undertook to complete the manuscript for publication in this form. The writing has been done at odd moments taken from other exacting duties, and no opportunity has been available for conference with his former associate in the field examination. The report has doubtless suffered as a result, although the general conclusions were formed during the field work and are believed to represent the opinions of both geologists.

The field work was done in the spring and summer of 1916, and the manuscript was begun the following fall. At that time the sections on copper and zinc deposits were written by Mr. Jones; the other sections have been written then and since by Mr. Umpleby.

---

[1] Ransome, F. L., and Calkins, F. C., The geology and ore deposits of the Coeur d'Alene district, Idaho: U. S. Geol. Survey Prof. Paper 62, 1908.

1

ACKNOWLEDGMENTS.

Acknowledgment of help is always difficult in a large mining district, and this is particularly true of the Coeur d'Alene district. Many eminent geologists and engineers have at different times studied parts of the district in great detail in connection with litigation, each leaving an imprint on the geologic thought of the miners and operators, so that it is quite impossible to give proper credit, even for observations of unusual interest. The work of Mr. Oscar H. Hershey stands out because of his detailed mapping of the Wardner area. His records are kept up to date and were fully accessible to the writers at all times during the field work. Mr. Rush White, from his intimate acquaintance with the Morning mine and many of the mines on Canyon Creek, rendered great assistance in the collection of evidence bearing on particular points. The mining companies, with the single exception of the Day interests, gave full access to their mines and complied cheerfully with every request for maps and other detailed records. As litigation was expected in connection with some of the mines, this is a particularly striking expression of the confidence of the operators in the Geological Survey. Although it is impossible to list all those who gave freely of their time and intimate knowledge of certain deposits, it would be ungrateful not to acknowledge in particular the courtesies of Mr. Stanly A. Easton, general manager of the Bunker Hill & Sullivan Co.; Mr. Frederick K. Burbidge, general manager of the Federal Mining & Smelting Co.; Mr. James F. McCarthy, manager of the Hecla mine; and Mr. Stanley R. Moore, superintendent of the Success mine.

LOCATION AND EXTENT OF AREA.

Shoshone County, Idaho, in which lies the well-known Coeur d'Alene district, comprises a large part of the narrow portion of the State that extends northward to the International boundary. (See Pl. I.) From the summit of the Continental Divide on the east the county extends west about halfway across the State. It lies within the greatly dissected plateau region of Idaho, which is characterized by deep valleys separated by high, timbered divides.

The accompanying map (Pl. I, in pocket) shows the essential topographic and geologic features of the area so far as they are known, but as most of the mines are situated within the area covered by the report of Ransome and Calkins, already cited, and as that report is no longer available the geologic map accompanying it is reproduced without modification in this report as Plate II (in pocket).

The main line of the Chicago, Milwaukee & St. Paul Railway crosses the southern part of the county, the Northern Pacific Railway has a branch line to Wallace from Missoula, Mont., and the Oregon Railroad & Navigation Co. has a branch to Wallace from Tekoa, Wash. A branch railroad has recently been extended up the valley of Pine Creek a few miles, affording transportation to the zinc mines of that area, and another follows the valley of North Fork of Coeur d'Alene River to the old gold-mining camp of Murray.

The county is about 75 miles in greatest length from north to south. At the north it is only about 12 miles wide, but the eastern border trends southeastward, and along its southern border it is nearly 60 miles wide.

HISTORY AND PRODUCTION.

Ore deposits were discovered in Shoshone County in 1878, but active exploitation began in 1884, during the period when a wave of prospectors from the gold fields of California spread over the Rocky Mountain States and discovered deposits in most of the dis·tricts that have since become prominent in the history of western mining development. The thoroughness of their search in Idaho is attested by the comparatively few districts that have been opened since 1890. Perhaps most of these were known to the early prospectors but because of the base character of the ores were not considered desirable at that time.

Gold was first discovered in the vicinity of Murray in 1883, and in the following year $250,000 worth was produced. It was only two years later when the lead-silver deposits in the vicinity of Wardner were found, and by 1889 these deposits were yielding annually ore containing 17,500 short tons of lead and approximately 1,000,000 ounces of silver. Since then neither the lead nor the silver output of the county has ever been less than this amount. During recent years about 12,000,000 ounces of silver and 175,000 short tons of lead have been produced annually. Copper came into prominence with the opening of the Snowstorm mine in 1904 and attained its maximum output of 9,000,000 pounds in 1908. Since then the production has gradually declined to about 2,500,000 pounds, and most of it comes from cupriferous lead-silver ores.

Zinc has long been known in the district, but it was first marketed in 1905, when 144,000 pounds was sold. Since then the annual production has rapidly increased.

The annual production of the district in the principal metals for each year since the discovery of the deposits appears in the following table:

Ore sold or treated in the Coeur d'Alene district, Shoshone County, Idaho, 1884–1920, number of producers, and content and total value of gold, silver, copper, lead, and zinc.[a]

Year.	Number of producers.	Ore (tons).	Gold.	Silver (fine ounces).	Copper (pounds).	Lead (short tons).	Zinc (spelter) (pounds).	Total value.
1884			$258,375					$258,375
1885			376,607					376,607
1886			182,371	116,246		1,500		436,335
1887			152,276	340,000		5,980		1,022,996
1888			211,867	554,000		8,000		1,438,227
1889			174,310	1,095,265		17,500		2,532,978
1890			165,360	1,499,663		27,500		4,132,506
1891			206,700	1,825,765		33,000		4,868,356
1892			227,370	1,195,904		27,839		3,538,684
1893			304,841	1,963,561		29,563		4,258,621
1894			362,365	2,343,314		30,000		3,816,026
1895			381,134	2,471,300		31,000		4,016,049
1896			359,017	3,163,657		37,250		4,703,971
1897			339,070	3,756,212		57,777		6,764,010
1898			268,937	3,521,982		56,339		6,565,287
1899			177,803	2,737,218		50,006		6,263,404
1900			118,935	5,261,417		81,535		10,588,707
1901			101,593	4,339,296		68,953		8,731,662
1902			98,410	5,033,928		74,739		8,847,552
1903			158,146	5,471,620		103,691		11,885,078
1904	27	1,410,245	46,015	6,143,001	1,424,400	112,584		13,592,014
1905	37	1,526,927	38,987	7,292,986	5,225,606	126,928	144,000	17,198,856
1906	36	1,622,975	86,607	7,944,338	6,393,940	126,011	2,054,998	21,133,963
1907	33	1,541,670	81,696	7,266,862	7,199,648	114,721	6,948,655	18,888,203
1908	43	1,551,680	80,167	6,364,552	9,042,876	102,069		13,220,853
1909	41	1,569,332	89,418	6,201,157	8,906,944	106,779	1,280,000	13,724,065
1910	39	1,639,781	65,076	6,703,080	6,018,688	109,879	5,526,717	14,416,910
1911	50	2,004,487	86,035	7,383,899	2,926,551	127,419	8,310,435	16,306,680
1912	52	2,108,037	84,430	7,558,314	4,386,403	123,276	13,800,181	18,313,604
1913	48	2,289,226	81,749	9,337,109	5,097,894	148,370	21,415,565	20,767,410
1914	38	2,152,268	64,157	12,178,194	4,242,662	169,849	41,523,383	22,728,903
1915	52	2,255,475	46,433	11,158,955	1,941,296	164,199	69,685,003	30,119,424
1916	63	2,516,325	46,459	11,639,841	2,370,610	178,117	86,238,283	44,424,716
1917	66	2,522,127	85,683	11,241,126	2,876,911	183,004	77,724,221	50,054,297
1918	52	1,918,052	245,450	8,447,219	2,706,535	139,307	43,661,314	33,115,903
1919	38	1,308,063	179,582	4,815,200	1,474,642	83,833	15,994,229	15,900,815
1920	39	1,822,488	121,900	6,386,663	571,119	118,105	27,932,326	28,347,791
			6,155,331	184,752,844	72,806,725	2,988,622	422,239,310	487,299,838

[a] 1884–1903, Ransome, F. L., U. S. Geol. Survey Prof. Paper 62, p. 82, 1908. 1904–1920, U. S. Geol. Survey Mineral Resources.

ORE RESERVES.

The ore reserves of the Coeur d'Alene district remain about the same as at the time of the examination by Ransome and Calkins. Several mines that were active when they visited the district have since been exhausted, but others of comparable importance have taken their place and new ore bodies have been found in several mines then active. The Standard-Mammoth, one of the most productive deposits of the district for many years, has been abandoned, and the Tiger-Poorman is no longer worked. Since 1904, however, a continuation of the Standard-Mammoth lode on the Green Hill-Cleveland property has been discovered and has yielded much ore. It is now apparently near exhaustion. During the last decade the Hecla mine has opened several new levels on three ore bodies that in the aggregate have increased in area with increasing depth. Very rich deposits have been found and exhausted in the Caledonia and Senator

CHARACTERISTIC TOPOGRAPHY OF SHOSHONE COUNTY, IDAHO.

View northward across valley of North Fork of Coeur d'Alene River above Mullan.   Photograph by R. A. Standow.

A. VALLEY OF MILO CREEK, SHOSHONE COUNTY, IDAHO.

Town of Kellogg in foreground.   Kellogg Peak in left distance.

B. SUMMIT TOPOGRAPHY AS SEEN FROM A POINT NEAR KELLOGG, IDAHO.

View looking north.

Stewart mines. It is not impossible, however, that continuations of these ore bodies will be discovered beyond the faults against which they appear to terminate. The Hercules ore body is said to be smaller on the lower levels, but this is probably offset by important discoveries in the Tamarack and Custer mines, owned by the same company. The Morning mine and the Bunker Hill and Sullivan mine have developed about as much ore on lower levels as on the upper ones, and both must be considered as holding a very promising future. Perhaps the most noteworthy discovery in the district during the period since the former report was issued is the great zinc body in the Interstate-Callahan mine, from which dividends of over $5,000,000 were paid in three years. Also, on Pine Creek, zinc bodies of unusual promise have been opened and only partly developed.

Ransome, in his summary of prospects for the district, based on his observations in 1904, expressed the opinion that the Bunker Hill and Sullivan mine might reasonably be expected to maintain its rate of production for 20 years or more. Existing conditions appear to warrant the forecast that at least 10 years will elapse before a noteworthy decrease in mining activity in the Coeur d'Alene district need be expected. It is likely, however, that zinc will gradually attain a more prominent relative position in the mineral output of the county—not only because the zinc deposits on Pine Creek are of great promise but also because a greater proportion of the zinc is being recovered than formerly and, as Ransome foresaw, several of the ore bodies become richer in zinc with increased depth.

The gold production of the district, which was a large element in its output during the first years after its discovery in 1883 but which gradually became of little relative importance, has recently taken on new life by reason of the installation of a gold dredge on Prichard Creek below Murray. The considerable extent of the gravel deposit at this place warrants the opinion that gold dredging will be continued for a number of years.

During the period of high prices for tungsten in 1915–16 a considerable quantity of this metal was produced, but from the spotty nature of the deposits it would seem that the district is not likely to become an important source of tungsten. Antimony ore also was produced during the World War, but no large continuous bodies of it have been recognized, although two or three mines may continue to yield a small amount for a number of years.

Copper mining has ceased to be important in the central district, but it is not improbable that the present production of about 2,500,000 pounds a year may be maintained for several years from outlying parts of the county.

In a rugged country heavily covered with timber and brush, such as Shoshone County, the probability of finding new deposits must be recognized in any attempt to forecast its future output. It is an eloquent argument as to the possibilities of the county that such deposits as the Interstate-Callahan, the Caledonia, and the Senator Stewart ore bodies, situated in the more frequented parts of the area, could have remained undiscovered for so many years. Much of the region is almost inaccessible, and away from the main streams there are few trails to encourage the prospector in exploration. Almost every part of the county and almost every rock formation within it is known at some place to be mineralized. Hence it is almost certain that new discoveries will be made in the future.

TOPOGRAPHY.

Shoshone County is a greatly dissected plateau standing at an elevation of about 6,000 feet in the southern part of the area and sloping gently northward to an elevation of about 5,500 feet in the northern part. Low points in the county are 2,142 feet where Coeur d'Alene River leaves it and probably about the same elevation along the lower part of St. Joe River. Wallace (Pl. III), the county seat, is 2,728 feet above sea level.

The main river valleys lie athwart the structural axes of the region and are bordered by steep slopes, which are heavily timbered except where the demands of mining have stripped them of their virgin growth or the ravages of forest fires have denuded them. In most places a dense second growth makes geologic observation extremely difficult and, combined with the rugged character of the country, limits travel to a few main highways and a few more trails.

The ridges of the area are even-crested or gently sloping, as shown in Plate IV, although from points of vantage low stretches in the sky line, as at Dobson Pass, suggest old river valleys long since abandoned by streams. Also in places, as along the South Fork of Coeur d'Alene River between Kellogg and Wallace, breaks in the contour of numerous secondary divides as they join the main valley suggest high terraces and if carefully studied might reveal the details of a complex physiographic history.

A few peaks rise conspicuously above the general summit level; perhaps the highest are the Three Sisters, at an elevation of 6,900 feet, and Goat Peak, 6,760 feet. In the central part of the county Kellogg Peak, near Wardner, is a landmark. (See Pl. V.)

The drainage of the area is entirely westward. The northern part of the county is drained by Coeur d'Alene River and its North and South forks; the southern part by St. Joe River, except for the extreme southern part, which is drained by tributaries of Clearwater River.

GENERAL GEOLOGY.

SEDIMENTARY ROCKS.

The sedimentary rocks of Shoshone County comprise slate, quartzite, calcareous shale, and strata of intermediate composition, known as the Belt series, all of Algonkian age. The series is many thousand feet in thickness, and at many horizons in it ripple marks, clearly indicating deposition in shallow water, are well preserved. The subdivision into formations is more or less a matter of convenience, as few sharp transitions occur in the sequence. In general, however, the subdivisions recognized in one part of the county may be used in other parts without serious difficulty. Some of the formations are known to vary greatly in thickness between the northern and southern parts of the area.

The following summary of stratigraphy is compiled largely from the numerous reports on the geology of various parts of the county by Calkins and others.[2]

PRICHARD FORMATION.

The Prichard, the lowest formation of the Belt series in Shoshone County, is best developed in the central and southwestern parts of the county, where it attains a thickness of at least 8,000 feet with the base not exposed. It consists of blue shale and slate with subordinate interbedded gray sandstone and quartzite. In the vicinity of Kellogg the formation may be subdivided into upper, middle, and lower members by reason of a massive bed of white and gray quartzite that occurs within the series. In the Pine Creek area this quartzite member is 200 feet thick, and farther south, in the vicinity of Clearwater River, a middle quartzite member, probably the same one, is 1,500 feet thick. Here the slates above and below the quartzite have been converted to mica schist and gneiss, a phenomenon attributed by Pardee[3] to contact metamorphism.

BURKE FORMATION.

Overlying the Prichard formation is a series of pale-tinted siliceous shale and sericitic flaggy quartzite which attains a thickness

[2] Ransome, F. L., and Calkins, F. C., The geology and ore deposits of the Coeur d'Alene district, Idaho: U. S. Geol. Survey Prof. Paper 62, 1908. Calkins, F. C., A geological reconnaissance in northern Idaho and northwestern Montana: U. S. Geol. Survey Bull. 384, 1909. Pardee, J. T., Geology and mineralization of the upper St. Joe River basin, Idaho: U. S. Geol. Survey Bull. 470, pp. 39–61, 1910. Calkins, F. C., and Jones, E. L., jr., Geology of the St. Joe-Clearwater region, Idaho: U. S. Geol. Survey Bull. 530, pp. 75–86, 1912. Calkins, F. C., and Jones, E. L., jr., U. S. Geol. Survey Bull. 540, pp. 167–211, 1913. Jones, E. L., jr., Reconnaissance of the Pine Creek district, Idaho: U. S. Geol. Survey Bull. 710, pp. 1–36, 1919. Hershey, O. H., Genesis of silver-lead ores in Wardner district, Idaho: Min. and Sci. Press, vol. 104, pp. 750–753, 786–790, 825–827, June 1, 8, 15, 1912.

[3] Pardee, J. T., op. cit., p. 44.

of 3,500 feet in the central part of the county but which appears to thin southward to about 1,000 feet in the Clearwater basin and here contains layers of mica schist. For convenience in structural mapping in the vicinity of Wardner, Hershey [4] has used a central purple-gray bed as the basis for a subdivision into upper and lower members.

REVETT QUARTZITE.

The Revett quartzite overlies the Burke formation and is similar to it in composition, but the individual beds are much more massive and more purely siliceous. The thickness of the formation ranges from 1,200 feet east of Wallace to 1,500 feet in the extreme southern part of the county and to 2,300 feet at Wardner.

ST. REGIS FORMATION.

The St. Regis formation, made up of quartzitic sandstone and shale, is characterized by peculiar purple and green colors in the central part of the county, but toward the south and east the coloration becomes less marked and the formation can not be traced with certainty, being difficultly separable from the underlying Revett quartzite and the overlying Newland formation. The lower limit is in most places uncertain, as there is almost perfect gradation between the thin-bedded upper layers of the St. Regis and the underlying massive members of the Revett. Upward the St. Regis grades into the thin-bedded shale of the Newland. The St. Regis as mapped is about 1,000 feet thick and has a comparatively small areal distribution in the county.

NEWLAND ("WALLACE") FORMATION.

The Newland, first described in the Coeur d'Alene district as the Wallace formation, crops out over about one-half of Shoshone County. It is the thickest formation above the Prichard, in many places from 4,000 to 6,000 feet of beds being exposed. As a whole it is made up of thin-bedded calcareous rocks. The basal beds are greenish and resemble the upper St. Regis in being very slightly calcareous. Above these beds the number of calcareous sandstones and shales increases to a horizon near the middle of the formation, where an abundance of calcareous sandstone incloses irregularly spaced beds of impure limestone. In the upper members of the formation layers of blue shale predominate. These differences may be used as the basis for recognizing upper, middle, and lower subdivisions of the formation. In the southern part of the county the formation has been metamorphosed over wide areas to mica schist, hornstone, quartzite, and related rocks.

---

[4] Hershey, O. H., op. cit., p. 2.

STRIPED PEAK FORMATION.

The Striped Peak is less widely distributed in Shoshone County than any other formation of the Belt series. It has been recognized in only a few places in the central part of the county. In lithologic character it is very similar to the St. Regis, being made up of lilac-colored quartzitic beds. Along the crest of the peak after which it is named about 1,000 feet of beds are exposed.

IGNEOUS ROCKS.

The igneous rocks of Shoshone County, with the exception of a few small patches of basalt in the lower drainage basin of St. Joe River, are all intrusive. They may be grouped broadly as monzonite, lamprophyre, and diabase. The monzonite masses are considered to be outliers of the Idaho batholith, which crops out in a continuous area of more than 24,000 square miles in the central and southern parts of the State. This great intrusive mass is younger than any of the Algonkian or Paleozoic rocks of the region and by several writers has been considered of late Mesozoic age. Within the borders of Shoshone County it has caused intense metamorphism in many places where it comes into contact with calcareous phases of the Belt series. In some places where no igneous rocks appear at the surface the development of contact-metamorphic minerals suggests the presence of granitic masses at no great depth.

The monzonite rocks are locally known as granite, but microscopic examination has shown that plagioclase feldspar enters largely into their composition. The amount of quartz present is very different in different places, so that locally the rock might be classified as quartz monzonite, granite, or syenite. On the whole, however, monzonite is the most applicable term for it.

Diabase occurs in several parts of the area, and the largest mass is known as the Wishards sill. This sill, which is traceable for more than 30 miles in the upper drainage basin of St. Joe River, is intruded near the middle of the Newland formation. It has been deformed along with the inclosing beds and is locally metamorphosed by the granitic intrusions, thus being clearly the oldest igneous rock of the region. Elsewhere in the county diabase occurs as narrow dikes, the rock of which can scarcely be distinguished megascopically from the lamprophyre in dikes that follow the same system of fractures. Some of the lamprophyres are definitely known to be younger than the ore deposits and must be considered differentiates from the granitic intrusions. It seems likely, therefore, that diabase younger than the monzonite and much younger than that forming the Wishards sill also occurs in the region.

In the southern part of the county dikes of granodiorite, porphyritic monzonite, and pegmatite have been reported. Anorthosite occurs in at least one locality.

The basalt shown in small areas on the geologic map (Pl. I) consists of remnants of a comparatively recent lava flow that at one time was possibly continuous with the basaltic lavas of the Columbia River valley.

The igneous rocks of the area have been carefully described in the several reports referred to in the first part of this chapter, and it is not considered necessary to repeat the detailed petrographic descriptions here. On the whole, igneous rocks constitute only a small proportion of the surface formations of Shoshone County. The monzonite has played so important a part, however, in the development of the ore deposits in this region that particular attention should be called to its geographic distribution. North of the South Fork of Coeur d'Alene River several masses of granitic rocks appear at the surface along the axis of an ill-defined anticline that extends north-northeast. Farther south, between the South Fork of Coeur d'Alene River and St. Joe River, there are only a few outcrops of granitic rocks, although prospectors have reported that an area of noteworthy size occurs west of the county line. South of the St. Joe, particularly along the divide between its tributaries and those of Clearwater River, monzonite is exposed prominently at the surface. Here there is a broad region of rather intense metamorphism which is characterized by the development of lime silicate minerals, particularly in beds of the Newland and Prichard formations. Pardee [5] has pointed out that the intensity of the metamorphism increases toward the west and south.

Adjacent to the northern intrusions Calkins observed metamorphism at several places, and a detailed description of the phenomena in the vicinity of the Success mine is given on pages 146–150.

STRUCTURE.

General features.—The structural features of Shoshone County are extremely complex and involve faulting and folding on a large scale. Previous publications accompanied by detailed maps record the more important features, although in parts of the county no fully successful attempt has ever been made to record on a geologic map the detailed structural relations. The lithologic character of the rocks is so unusually similar through hundreds and even thousands of feet of section that the recognition of faults of less than a few hundred feet displacement is in most places impossible.

---

[5] Pardee, J. T., op. cit., p. 49.

All the strata are folded, and over wide areas the average inclination of the once horizontal beds is in excess of 45°. The northwestern and southeastern parts of the county comprise areas of much less folding than a north-northeast zone that separates them. It is along this zone that the oldest rocks appear at the surface, indicating that although much broken by cross faults and igneous intrusions, the zone represents a broad anticlinal arch. Toward the south it seems to bend eastward, but here there has not been sufficient mapping toward the west to make the broader relations clear.

The most noteworthy structural feature of the county is an east-west fault or fault zone, known as the Osburn fault, that extends across the entire county and for many miles beyond both its eastern and western borders. About 6 miles south of this is another fault that crosses the county and is known as the Placer Creek fault.

Osburn fault.—The Osburn fault, long recognized as an earth fracture of major importance, is the most pronounced tectonic feature of the region. The fault crosses the county in a west-north-westerly direction. It can be readily recognized as far west as the mouth of Fourth of July Canyon, and it continues eastward beyond the Continental Divide for a long distance. The geology on opposite sides of the fault has been studied in detail for a distance of about 30 miles, both on the surface and in numerous mine workings. Throughout this distance it approximately coincides with the old valley described by Calkins and Jones [6] as extending from Spokane, Wash., to Deer Lodge, Mont., a distance of 300 miles.

At Mullan the fault disappears beneath the alluvial floor of the valley, which is so broad for some distance eastward as to baffle any effort to trace the fault directly, and the upper part of the valley, which is narrower, has not been fully examined for evidence of the continuation of the fault to the east. The old valley, however, is clearly located on a fault or fault zone whose identity with the Osburn fault, though it can not be absolutely proved, is strongly indicated by its direction and position and the fact that, like the Osburn fault, it effects a downthrow on the south.

The fault has been crossed in numerous mine workings and prospects, and thus an excellent opportunity has been afforded to observe it in detail. In general there is a central mass of gouge a few feet wide, with numerous parallel gouge seams in both the hanging wall and the footwall. Thus the fault must be considered a complex fracture with many planes of movement distributed through a zone locally 100 or 200 feet wide. The fault plane in every exposure examined dips steeply south. Locally igneous material, principally lamprophyre, follows the fault zone, indicating

[6] Calkins. F. C., and Jones, E. L., jr., Economic geology of the region around Mullan, Idaho, and Saltese, Mont.: U. S. Geol. Survey Bull. 540, pp. 9, 16, 1913.

that it existed as a zone of weakness during the period of magmatic activity in this region.

The apparent displacement caused by the Osburn fault is widely different in different places, ranging from less than 1,000 feet to more than 10,000 feet, but even after allowing for such vertical displacements it is impossible to fit the opposite sides together into anything like a regular arrangement. Anticlines are opposite synclines; major north-south faults are cut off abruptly; and cross folds that might explain the great differences in apparent displacement are lacking. These difficulties suggested a pronounced horizontal component of movement and led Hershey[7] to make a detailed study of the problem. He found that a peculiar purple phase of the Burke formation, abundant in the vicinity of Kellogg south of the fault, was absent from the Burke formation directly north of the fault but that 12 to 15 miles farther east, in the vicinity of the town of Burke, this same purple banding characterizes the formation. The purple coloring in each locality increases in intensity eastward. Hershey also correlates several minor structural features north and south of the fault, each indicating a movement of the south side from 10 to 16 miles westward relative to the north wall. Even more convincing than these evidences, however, are certain broad features brought out by the writers' work in the Pine Creek district in 1916. Here the rocks have a domelike structure, with the Burke and Revett formations flanking the Prichard on the east. The continuation of this structure north of the fault appears between Osburn and Ninemile Creek. Also the lead-silver deposits at Wardner lie east of the Pine Creek zinc-producing belt, and north of the fault the Canyon Creek lead-silver deposits are about the same distance east of the zinc deposits on Ninemile Creek. A large fault in the Pine Creek area, known as the Government Gulch fault, may be correlated with some assurance with the Carpenter Gulch fault, which dips in the same direction and also forms the contact between Burke and Prichard rocks. As some of the offset structural features dip east and some of them dip west, yet all of them indicate a shift to the west, the displacement must be considered an actual westward shift of the south side relative to the north side. This new evidence indicates a horizontal component of movement of about 12 miles, which accords with Hershey's conclusion that the horizontal component " may have been 10 or 15 miles."

In view of the great horizontal component of movement, curvature of the fault plane suggests problems of particular interest. If the points where the fault leaves the east and west boundaries of the area represented on the map of the Coeur d'Alene district (Pl. II)

---

[7] Hershey, O. H., Origin and distribution of ore in the Coeur d'Alene (private publication), Min. and Sci. Press, 1916.

are used to determine a straight line, it appears, after corrections for dip of the fault plane and differences in elevation of the surface, that the fault curves gradually southward, reaching a point 4,200 feet south of its general course in the vicinity of Wallace. Thus as movement of the hanging wall relatively westward progressed there was near Wallace a protuberant part of the footwall offering particular resistance to the movement. The net tendency of the stresses would be to slice the protruding segment parallel to the general course of the Osburn fault. Thus may be explained the localization and intensity of the east-west faulting in the area between Burke and Mullan and westward beyond Ninemile Creek. But what of the hanging-wall segment opposite this bulge in the footwall? As it moved westward it had a tendency to occupy positions successively farther from the footwall. But lateral pressure was sufficient to keep the fault fissure closed, and hence there must have been a crowding forward of this part of the hanging wall to keep in con-

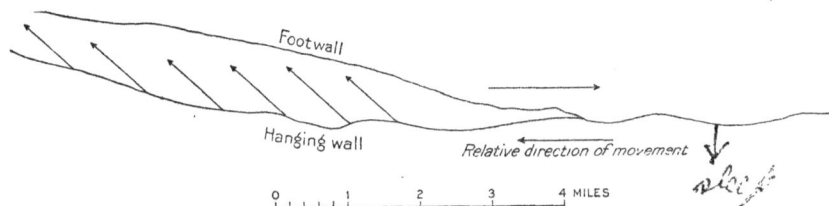

FIGURE 1.—Diagram illustrating the effect of horizontal displacement along the curved plane of the Osburn fault.

tact with a footwall that curved northward away from it. Furthermore, as the net result along the Osburn fault was a relative shifting of the south side westward, there must have been a strong horizontal component of force in that direction. This, however, was probably the major component of a stress that operated obliquely to the direction of movement. Thus it is probable that as the shifting progressed planes of fracture oblique to the main one would develop, breaking the rocks into slices, and that each slice would advance more than the one northeast of it. In this way the intense slicing of the Wardner area and the progressive shifting of the south slices westward along lines oblique to the Osburn fault may be accounted for, as illustrated by figure 1.

The peculiar localization of particularly intense faulting in the Wardner area south of the fault and in the area roughly bounded by Mullan, Burke, and Ninemile Creek north of the fault, whereas simple structure is found opposite both of these areas, is a striking feature of the region and baffling to explain, except as a direct result of the horizontal movement along the curved plane of this great displacement in the earth's crust.

Placer Creek fault.—The Placer Creek fault is also one of great persistence. It crosses the county in an east-west direction and is known to cross the Continental Divide. The displacement along this fault, however, does not seem to be nearly as large as that along the Osburn fault, although the rocks on opposite sides of it have not been studied in detail. In general the structural features north and south of it are in rough alinement, and throughout most of its course it is bordered by rocks of the same formation. The fault, however, has a marked topographic expression, forming low saddles that may be seen for miles from points of vantage and that establish its essential continuity. Its east end is known as the Roland fault.

Other faults.—The numerous other large faults of this region have been described in detail in previous papers, and as the recent investigation has added nothing of interest concerning them the description will not be repeated here. It should perhaps be pointed out, however, that the New Era fault, which Hershey [8] maps as cutting the Osburn fault, has been shown by more recent development in the Stewart mine to be cut off by the Osburn fault, which is therefore, so far as known, an uninterrupted fracture.

ORE DEPOSITS.

GENERAL CHARACTER.

The mineral deposits of Shoshone County yield ores of lead, silver, zinc, copper, gold, antimony, and tungsten. Most of the output has come from the lead-silver deposits, which have been increasingly productive since their first exploitation, more than three decades ago. Gold was discovered in 1878 and was of prime importance during the first decade of mining in the area, which began about 1884. During a later period copper yielded a large revenue, but now almost no copper is produced except as a minor product from the lead-silver ores. Recently large zinc mines have been opened, and numerous promising prospects indicate that this metal should continue for many years as an important factor in the mineral production of the county. Under favorable market conditions antimony and tungsten are mined in comparatively small amounts.

The deposits of the area fall into two general groups according to whether the ore-depositing solutions filled preexisting spaces or made space for themselves by metasomatic replacement of the wall rock. It is of particular interest that in an area of quartzite and siliceous slate, rocks commonly found to be relatively resistant to replacement, metasomatic processes should have played the dominant part in the formation of ore bodies. The principal exceptions are

---

[8] Hershey, O. H., Genesis of silver-lead ores in Wardner district, Idaho: Min. and Sci. Press, June 1, 1912.

the gold-quartz veins near Murray, which fill fissures in Prichard slate and follow its bedding planes, forming the bed veins described in detail by Ransome.[9] Certain of the deposits near Wardner also occupy fissures that may have stood open at the time of deposition, although here the marked irregularity of the walls clearly indicates that replacement took part in their development.

The deposits belong to the comparatively rare group of lodes in which siderite is the dominant gangue mineral but are different from other known members of the group, as they are inclosed principally in quartzite. Deposits closely resembling them occur in the Siegerland district of Westphalia, where lodes as much as 20 meters wide traverse arenaceous clay slate, sandstone, and graywacke of Devonian age. The ore consists of siderite, quartz, sphalerite, galena, and small amounts of copper and cobalt minerals, deposited in the order named.[10] Siderite is a gangue mineral in the deposits of Leadville and Redcliff, Colo., and Pioche, Nev., the Iglesias lodes in Sardinia, sparingly in the Baudwin mines of Burma, and in the Dome, Texas, and Wood River districts, Idaho. In most of these districts the siderite was formed early in the general period of ore development and was later fractured and replaced by sulphides, as discussed on pages 135–138.

Throughout the lead mines of the county a definite sequence of mineral deposition is recorded, the succession being siderite, quartz, sphalerite, galena. The numerous siderite lodes (pp. 31–32) are considered as representing an early condition of most of the other lodes, and the history of the zinc deposits is believed to be essentially similar to that of the lead deposits up to the time when lead was introduced. It is recognized, however, that in different parts of the area the different stages of deposition varied in relative importance, so that by no means all the lead deposits would have made zinc mines at one stage in their history. On the other hand, the prevalence of sphalerite as rounded cores in galena and the replacement contacts between galena and the older minerals make it clear that the stage of lead deposition was accompanied by the removal of considerable material that had been earlier deposited in the lodes.

DISTRIBUTION.

GEOGRAPHIC DISTRIBUTION.

The mineral deposits of known importance in Shoshone County are grouped in a comparatively small area in the drainage basin of the South Fork of Coeur d'Alene River. North of the river are the

---

[9] Ransome, F. L., and Calkins, F. C., op. cit., p. 141.
[10] Beyschlag, F., Krusch, P., and Vogt, J. H. L., Ore deposits (translated by S. J. Truscott), vol. 2, p. 800, London, 1916.

large mines near Mullan and Burke and on Ninemile Creek; south of it are the Wardner and Pine Creek deposits. Numerous prospects are widely distributed over the areas north and south of the principal deposits, but none of them have proved to be important sources of production. North of the principal mineral area are the Empire Copper, Jack Waite, and numerous mines east of Murray; to the south are several groups of deposits in the St. Joe basin described by Pardee and by Calkins and Jones in the reports already cited. Thus mineralization is known to have extended sporadically over most of the county.

Although the mineral deposits are widely distributed, the different metals distinctly favor certain broad areas. Gold deposits occur principally in the vicinity of Murray and in a small area east of Kellogg. Copper is disseminated in beds east of Mullan and occurs in lodes over a broad area north of St. Joe River and east of Avery. Other localities are on Sliderock Mountain, Black Prince Creek, and the Little North Fork of Coeur d'Alene River. The principal deposits of lead ore are near Wardner and in an area embracing Mullan and the upper drainage basin of Ninemile Creek. Lead deposits of secondary importance occur in the general vicinity of Murray, along a strip between Wardner and Wallace, on Slate Creek, and on Pine Creek. Zinc is closely related to lead in distribution but in general overlaps the lead area from the west. The most conspicuous illustration of this overlap is in the Morning mine, near the eastern limit of the predominantly lead-bearing area. Here the west end of the lode is much higher in zinc than the east end, although the deposit as a whole is more valuable for lead than for zinc.

GEOLOGIC DISTRIBUTION.

Mineral deposits occur in all the rock formations of the county, but without exception all the large mines are situated in the areas of more marked structural disturbance. The large mines south of the South Fork of Coeur d'Alene River form a group about 15 miles west of those north of the river. They are also south of the Osburn fault, which, from several distinct lines of evidence, is believed to represent a horizontal displacement of comparable extent in the same direction. Between 2 and 3 miles still farther south is the Placer Creek fault, also with southward dip, which extends entirely across the county. The marked curvature of this fault seems to preclude a horizontal component of movement equal to that of the Osburn fault, although its known extent of more than 40 miles proves it to be an earth fracture of great magnitude. Innumerable minor displacements, doubtless related to these major ones in origin, occur in an east-west zone, north and south of which the rocks are much less disturbed and individual formations are continuous over wide areas, as may be seen from Plate I. It is particularly noteworthy that the great deposits of ore are confined to the general zone of most pronounced structural disturbance. The ore bodies, however, do not follow the major faults but distinctly favor minor fractures, few of which represent sufficient displacement to cause a noteworthy offset of formations at the surface.

The deposits show a further grouping with respect to structure. A broad anticline, much older than the east-west faults, crosses the productive part of the area in a northeasterly direction. It is shown by a central belt of Prichard rocks, conspicuously interrupted by cross faulting, which is bordered by younger rocks both to the northwest and to the southeast. The principal ore bodies occur on the eastern flank of this anticline, in the general zone represented by the upper Prichard, Burke, and Revett rocks but locally extending into the Newland formation, as at the Gold Hunter mine. A narrow belt east of the axis of the anticline embraces the several areas of granitic rock which occur in the central part of the county. The principal mines, however, are not coextensive in distribution with the area of igneous intrusion but rather are near the southern member of the chain of granite masses, and if the Wardner area were shifted back to its original position near Wallace they would be well to the south of it.

On combining the features of structural control described above it appears that the principal ore deposits of the county are roughly grouped about the intersection of an axis of uplift and igneous intrusion with a broad zone of remarkably extensive faulting. The relations are further discussed in the sections on geologic structure and ore genesis (pp. 10–14, 141–150).

The principal ore bodies are inclosed in the Prichard (upper part), Burke, and Revett formations, although ore has been deposited in all the rock formations of the area. In general the gold and zinc deposits are in the Prichard, the lead-silver deposits in the Burke and Revett, and the copper deposits in the Revett. Among the zinc deposits the Interstate-Callahan, Rex, Success, and the lodes on Pine Creek are in the Prichard formation, but the Helena-Frisco is in the Burke, and the Morning, a lead mine but still an important source of zinc, is in the Revett. The lead deposits are distributed through an even wider stratigraphic range. The Hypotheek, Jack Waite, Paragon, and Monarch, the upper levels of the Callahan, Rex, and Success, and probably the lower levels of the Hercules are inclosed in Prichard slate or quartzite. Inclosed in Burke quartzite are the Standard-Mammoth, Greenhill-Cleveland, Tiger Poorman, and Hecla, and parts of the Bunker Hill and Sullivan. The Revett quartzite is the country rock of the Morning lode and most of the great lodes near Wardner. The principal lead deposit in the St. Regis is the Carbonate Hill, and in the Newland the Gold Hunter. The copper deposits of the Snowstorm belt are all in the Revett quartzite, but toward the south copper is found widely in the Newland formation and locally, as at the Carney copper mine, in the St. Regis.

No deposits of large size are known in areas of granitic rocks, although enough sulphide has been deposited within granitic walls to indicate the age relation of the two phenomena. The Success, Frisco, Sunset, Lily, American, and Nellie Pollar deposits all show sulphides deposited along fractures and as irregular replacement bodies in monzonite or syenite. In some of them lime silicates are intimately associated with the sulphide minerals.

GENERAL FEATURES OF THE ORES.

The ores of Shoshone County yield lead, silver, zinc, and copper, together with minor amounts of gold, antimony, and tungsten. Silver is invariably associated with the galena, although a silver content greater than about one-third of an ounce to 1 per cent of lead usually indicates the presence of tetrahedrite. Galena, sphalerite, siderite, and quartz are present in most of the deposits but in widely variable relative amounts. The great ore body of the Bunker Hill and Sullivan mine consists of fine-grained galena, in places laminated as a result of shearing, distributed through siderite in interlacing seams and irregular replacement bunches. Sphalerite and quartz are present in very subordinate amounts, and tetrahedrite is rarely visible. In the Morning mine an ore of similar appearance contains about equal amounts of sphalerite and galena, and the Hecla ore, which is valuable only for lead and silver, contains quartz almost to the total exclusion of siderite. The prevalence of sphalerite in numerous deposits is illustrated by the Interstate-Callahan lode, which yields about four times as much zinc as lead and is low in silver and in which quartz is the principal gangue mineral. Stibnite has been detected in a few of the lead-silver lodes but attains commercial importance only in deposits in which other sulphides are scarce. Scheelite is confined to the gold veins. The copper ores consist either of bornite, chalcocite, and chalcopyrite disseminated in a coarse-textured bed of Revett quartzite or of mixed sulphides which occur in lodes inclosed in various formations of the Belt series. Only the former have been worked extensively.

The magnetic sulphide of iron, pyrrhotite, is common in the ores of the area, particularly in the zinc ore from Pine Creek and in some of the lead-silver ores of Canyon Creek. Closely associated with it in several places is magnetite, although this mineral attains its maximum development in the Success and Helena-Frisco mines, where lime silicates are abundant. Most of the ores contain pyrite.

In general, the ores of the county are fine grained, and the minerals are intimately mixed. For this reason losses in milling have been rather high, although the recent introduction of flotation in most of the plants has greatly increased the percentage of recovery. Most of the lead ores, as sent to the mills, contain from 8 to 11 per cent of lead and 3 to 12 ounces of silver to the ton, although several of the companies save a small percentage of crude ore of shipping grade. In places, as in the Yankee Boy and in parts of the Caledonia deposits, the ore is much richer in silver than elsewhere and tetrahedrite is abundant. The zinc ores, as mined in 1916, show a much wider range in tenor than the lead-silver ores. At that time the Interstate-Callahan was milling ore containing about 25 per cent of zinc, 6 per cent of lead, and 2 ounces of silver to the ton, which contrasts notably with ore from some of the other mines containing 8 per cent of zinc and about 2 per cent of lead. The copper ores of the area are principally of low grade, but their highly siliceous character, together with a relatively large amount of gold and silver, has made them desirable to the smelters. Much of the Snowstorm ore contained 4 per cent of copper and 6 ounces of silver and 0.1 ounce of gold to the ton, although its average yield was lower. The National ore is of much lower grade. The mining of antimony and tungsten has involved hand sorting to so great an extent that no adequate information is available as to the tenor of the ores. Both the stibnite and scheelite occur in bunches almost free from gangue or other minerals.

The relation between the several types of ore is one of relative importance of the major constituents and the abundance, meager development, or absence of the minor constituents. By contrasting extreme types an essential difference in genesis is suggested, but, as concluded by Ransome [11] in the following quotation, this suggestion is probably erroneous:

The contrast between such a deposit as that of the Granite [now the Success] mine, with its characteristic contact-metamorphic features, and the galena-siderite bodies of the Bunker Hill and Sullivan mine is striking and at first glance suggests that the two kinds of deposit are genetically dist'nct. Yet careful study of the whole district modifies this first impression. The ores of the Custer and Helena-Frisco mines, which are also in part associated with metamorphic silicates, supply a link between the ores of the Granite and Sixteen-to-One [now the Rex] mines and those of the Canyon Creek mines, while the occurrence of pyrrhotite in the Tiger-Poorman and Standard-Mam-

---

[11] Op. cit., p. 109.

moth mines and in the Highland Ch'ef prospect south of Wardner, as well as the presence of magnetite in the Morning ore, suggests that sharp mineralogical distinction between the ores of Ninemile Creek and those of Wardner would be difficult, if not impossible.

Even the gold-quartz veins, which are the most distinct group of deposits in the county, contain as accessory constituents all the important ore minerals found in the lead, zinc, and copper deposits. Variations from them to other types, however, are much more abrupt than between the other types. Between the gold ores and the copper ores a gradation may be recognized, as several veins near Kellogg and in the southern copper area may be considered equally valuable for each metal. On the other hand, the abundance of copper in the Caledonia and Black Bear Fraction deposits suggests a further transition to the lead-silver lodes. Also some of the lead deposits, such as the Terrible Edith, containing about $2 to the ton in gold, suggest a gradation directly to the lead-silver ores. The general aspect of the Terrible Edith, however, is so wholly different from that of the Golden Chest vein, near by, that doubt is at once cast upon the significance of gold in determining a genetic relation between this deposit and the gold veins. Possibly those veins should be considered a genetically distinct element in the mineralization of the county, as suggested by Ransome,[12] but of this there is abundant room for doubt. Gradation to the lead-silver deposits may be represented by the copper-gold lodes.

Oxidized ore has long since ceased to play an important part in the metal production of the county, although some still comes from the Caledonia, Sierra Nevada, and Hypotheek mines. Cerusite and native silver are the principal valuable constituents. The common gangue is porous limonite, bound together by a honeycomb of quartz and cerusite crystals and containing many vugs lined with these minerals, together with scales and wires of native silver and locally native copper, malachite, and azurite.

MINERALOGY.

The mineralogy of the ore deposits of the Coeur d'Alene district has been discussed in detail by Ransome,[13] and it is only necessary here to give a digest of his observations supplemented by data revealed by more recent development and the extension of the studies to include other parts of the county. The common rock-forming minerals are not mentioned except as they bear a definite relation to ore deposition.

---

[12] Op. cit., p. 143.
[13] Op. cit., pp. 90–103.

In all, 45 mineral species have been recognized in the ores of the county. Of these 13 are primary metallic minerals, 19 are primary constituents of the gangue, and 13 have resulted from processes of oxidation. The quantitatively important constituents of the ore, however, are much fewer in number. Four sulphides—galena, sphalerite, pyrite, and chalcopyrite—in widely different relative amounts in different places, occurring in a gangue of siderite and quartz, characterize most of the deposits. Locally barite is important as a constituent of the gangue, and in places tetrahedrite and stibnite are abundant. Lime silicate minerals occur only here and there. In the oxidized zone cerusite and limonite are abundantly developed.

The essential features of the mineralogy and some idea of the relations and distribution of the individual species are shown in the following tabular summary, in which the letters indicate the occurrence of the mineral—D, dominantly; A, abundantly; S, sparingly; R, rarely.

Ore minerals of Shoshone County, Idaho.

Mineral.	Composition.	Contact deposits. Primary.	Contact deposits. Secondary.	Lead-silver deposits. Primary.	Lead-silver deposits. Secondary.	Zinc lodes. Primary.	Zinc lodes. Secondary.	Copper deposits. Primary.	Copper deposits. Secondary.	Gold veins. Primary.	Gold veins. Secondary.	Antimony deposits. Primary.	Antimony deposits. Secondary.	Remarks.
Actinolite	CaO.2(MgO,FeO).4SiO$_2$	R												Intergrown with magnetite in Frisco ore.
Amphibole	Variable silicate.	A												Success mine.
Anglesite	PbSO$_4$				R									One specimen from Hypotheek mine.
Anthophyllite	(Mg,Fe)SiO$_3$			R										Intergrown with galena in Hercules ore.
Augite	Complex silicate								A					
Azurite	2CuCO$_3$.Cu(OH)$_2$				R	S	R		A		R			In mines near Mullan.
Barite	BaSO$_4$			A		S								
Biotite	HK$_2$(Mg,Fe)Al(SiO$_4$)$_3$													
Bornite	Cu$_3$FeS$_3$						?	S						
Calamine	(ZnOH)$_2$SiO$_3$						S		S					Probably occurs but not observed.
Calcite	CaCO$_3$	R		S		S		D	R	S	R	S		Evolution, Hypotheek, Constitution, etc.
Cerusite	PbCO$_3$				D			S	A					
Chalcocite	Cu$_2$S			S	R	S			A					
Chalcopyrite	CuFeS$_2$	R		S		S		S		S				
Chlorite	Complex silicate													
Copper	Cu				S			R	S					
Covellite	CuS				R				R					
Cuprite	Cu$_2$O								S					
Diopside	MgCa(SiO$_3$)$_2$	A												Caledonia mine.
Dolomite	(Ca,Mg)CO$_3$			R				R						Caledonia and Snowstorm mines.
Epidote	Complex silicate	A												Snowstorm mine.
Galena	PbS			D		A		S			A			
Garnet	Complex silicate	A								D		D		
Gold	Au			R	R					D				
Hedenbergite	CaFe(SiO$_2$)$_2$	A												
Hornblende	Complex silicate	A												
Limonite	2Fe$_2$O$_3$.3H$_2$O		D	S	D	S	D	S	D	R	A		D	Success mine.
Magnetite	Fe$_3$O$_4$	A			R				A	R				
Malachite	CuCO$_3$.Cu(OH)$_2$				R				A		A			
Massicot	PbO				R									
Platnerite	PbO$_2$				R									
Proustite	Ag$_3$AsS$_3$			R		A		A		A		A		Yankee Boy mine.
Pyrite	FeS$_2$	A		A		R		A		R		R		
Pyromorphite	Pb$_4$(PbCl)(PO$_4$)$_3$			S		A		A		S		A		
Pyrrhotite	Fe$_n$S$_{n+1}$	S				A				D				
Quartz	SiO$_2$	D		A		A		A		D		A		
Scapolite	Complex silicate	A												
Sericite	H$_2$Al$_3$(SiO$_4$)$_3$	A		A		A		A		S		S		St. Joe area.
Scheelite	CaWO$_4$	A		A		A		A		S		S		Locally in lodes; widely in country rock,

								Remarks
Siderite........	FeCO₃.....	R	D		D	A		A	Most abundant gangue mineral.
Silver..........	Ag........		R						Caledonia mine.
Specularite.....	Fe₂O₃.....	R							
Smithsonite.....	ZnCO₃.....				?				Not observed but probably present.
Sphalerite......	ZnS.......	A	A		D	S			
Stibnite........	Sb₂S₃.....		R			S	S	D	
Tetrahedrite....	Cu₈Sb₂S₇..		S		S				Usually indicates silver-rich ores.
Tourmaline......	Complex silicate.	R	R						Rarely in ore; widely in sediments.

CLASSIFICATION.

Ore deposits may be classified in many ways, according to the object in view. To the student of ore genesis a genetic classification has most advantages, and even in writing for the general reader there is much justification for grouping the deposits genetically. The most important problem in the study of ore deposits is to determine their origin, for a knowledge of the genesis of deposits of different types and their consequent characteristics is vitally necessary to intelligent exploitation, particularly in the early stages of development, and to the search for new deposits. Thus a genetic classification, because it contributes most to our knowledge of ore deposits, has noteworthy advantages, but in discussing a small area in which few types are represented close adherence to such a grouping may eliminate subdivisions which would elucidate the presentation. On the other hand, a classification to be most helpful to the miner, the metallurgist, and the average investor in mining properties must take into careful account the form and substance of the ore bodies and the nature of the inclosing rock.

In the later discussion the deposits are grouped by the kind of inclosing rock with primary subdivisions by genesis and form, and secondary subdivisions by substance. The major grouping is not followed out consistently, for to separate deposits inclosed in slate, siliceous slate, and quartzite would involve unwarranted repetition. It serves the useful purpose, however, of bringing together those deposits which show definitely the relative age of the ore deposits and the granitic intrusions.

Classification of the ore deposits of Shoshone County.

Deposits in igneous rocks.
Deposits in sedimentary rocks:
    Disseminated deposits:
        Copper deposits.
        Sphalerite and galena deposits.
        Pyrite deposits.
        Siderite deposits.
    Contact-metamorphic deposits.
    Replacement deposits along fissures and fracture zones:
        Lead-silver deposits.
        Zinc deposits.
        Copper deposits.
        Antimony deposits.
        Siderite deposits.
    Fissure fillings:
        Gold veins.
        Tungsten veins.

DEPOSITS IN IGNEOUS ROCKS.

No large deposits of ore occur in igneous rocks in Shoshone County, although several prospects in such rocks are noteworthy because of their significance in showing the age relation between the mineralization and the igneous activity of the region. The Success and the Frisco are the only large deposits that show any replacement of igneous rock, and both of these are predominantly in sedimentary beds. Locally in the Success mine apophyses from the monzonite are traversed by the ore, as for example between stations 300 and 308 on the 300-foot level (fig. 2), where a 10-foot dike nearly at right angles to the course of the ore body is exposed in opposite sides of a stope from which ore was extracted to a width of about 12 feet. Another equally clear example of ore cutting the monzonite appears on the same level near station 324, where an offshoot from the main ore body replaces a 4-foot dike and extends into the quartzite beyond.

The Frisco lode extends westward nearly to the igneous contact, and although the main lode does not reach the monzonite on levels that were accessible in 1916, veinlets of sphalerite together with much disseminated pyrite and less sphalerite were observed in the main monzonite mass on the 1,600-foot level west.

In the American prospect, immediately west of the Success mine, quartz and galena veinlets traverse monzonite in which sericite, chlorite, and pyrite are developed. Relations similar to these but more like those in the Success mine occur in the Clark mine, on Sunset Peak, and in the Lily and Nellie Pollar prospects, east of Murray. The ore at the Lily prospect, consisting of galena with associated chalcopyrite, has been formed by the replacement of syenite (Pl. VII, p. 41) adjacent to a 2-inch clay seam which separates it from Prichard slate. The bunches and veinlets of sulphides are most numerous next to the gouge seam, but they are also present, though in diminishing numbers, to a distance of 20 feet out into the syenite, as shown in two crosscuts 65 feet apart.

The Hecla lode on megascopic inspection seems to replace a lamprophyre dike, but microscopic observations fail to support this view. Ore and dike follow the same fracture zone, and in places the dike's irregular border suggests that it is replaced by ore minerals. Microscopic examination of specimens from such places, however, as Ransome [14] has stated, shows the contact to be an intrusive one with marginal chilling of the igneous material, minute dikelets of which extend into the ore. It is also noteworthy that none of the numerous ore seams that cut across the quartzite partings in the lode traverse the minette dike, and many of them may be seen to terminate abruptly against it. On the other hand in the Frisco mine the dike

[14] Op. cit., p. 176.

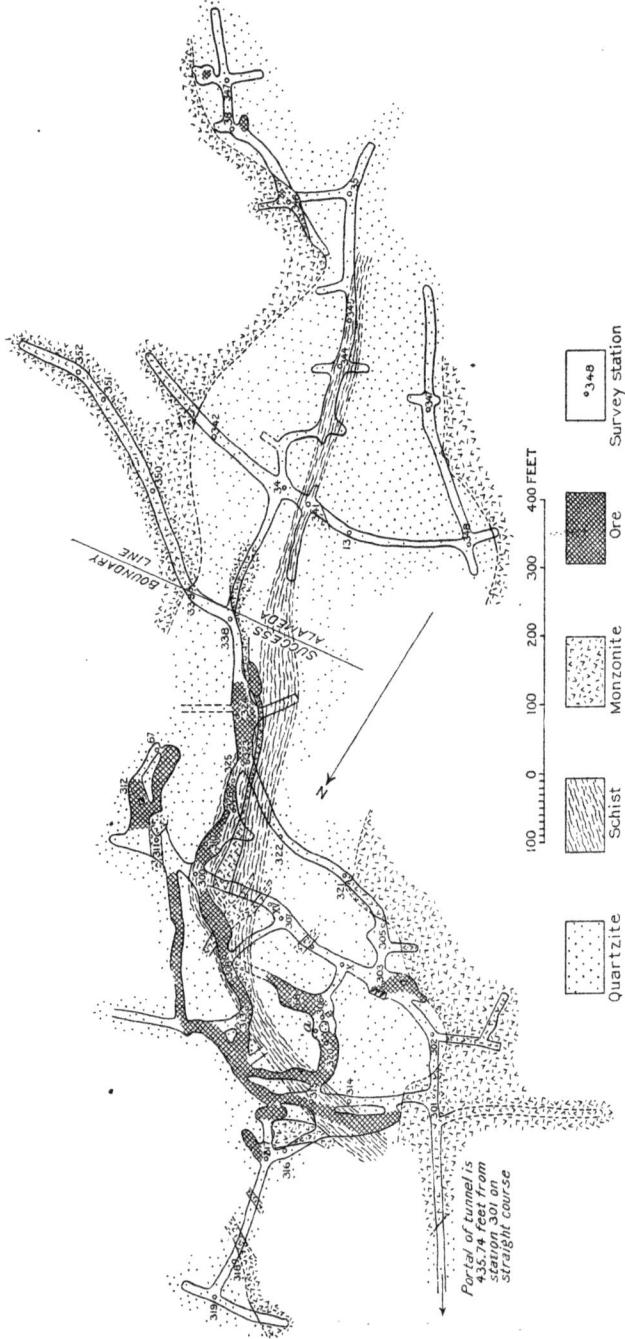

FIGURE 2.—Geologic map of the 300-foot level of the Success mine.

clearly cuts across the vein instead of following the fissure previously followed by the ore-depositing solutions.

These dikes are of particular significance in considering the genesis of the deposits, because they are clearly younger than the main granitic intrusions and represent late phases of the magmatic activity. In age, therefore, the ore deposits are closely related to the principal igneous rocks of the region.

In the southeastern part of the county siderite veins carrying chalcopyrite cut the diabase of the Wishard sill, but as this diabase is metamorphosed by the monzonite intrusion it represents an earlier period of igneous activity, and the occurrences can not be considered an exception to the above statement.

DEPOSITS IN SEDIMENTARY ROCKS.

The ore deposits of Shoshone County are almost unique among mineral deposits in that they are characteristically replacement deposits in quartzite. The Burke and Revett formations, which include the purest quartzites in the region, contain the principal lodes, although recently large bodies of zinc ore have been found in Prichard quartzitic slate.

The mineral occurrences are diverse in character, and no sharp lines of demarkation can be drawn between the several types. For purposes of systematic discussion, however, they are grouped as disseminated deposits, contact-metamorphic deposits, replacement deposits along fissures and fractures, and fissure fillings. Each of these groups is further subdivided according to the metals for which they are worked. In reading the following sections, therefore, it should be kept in mind that the grouping of the deposits is merely a breaking up of a complex whole in order to facilitate discussion.

Disseminated Deposits.

Ore minerals are widely disseminated in the rocks of Shoshone County, as might be expected in an area in which replacement of quartzite has played the dominant part in the formation of immense ore bodies. It is probable that all the minerals found in the lodes occur at some place in disseminated form, although only the copper minerals have yielded disseminated deposits of commercial value. Galena and sphalerite occur thus in the east end of the Bunker Hill and Sullivan mine and in narrow zones along ore bodies elsewhere in the district. Pyrite is widely disseminated in places, as illustrated in Plate VIII (p. 52). The most abundant and widely distributed mineral of all, however, is siderite, the characteristic gangue mineral of the deposits. This occurs in all formations but is most abundant in the sericitic quartzite in areas of pronounced folding and faulting.

Certain other minerals, considered in the section on metamorphism (pp. 38–41), which occur in clusters and isolated crystals, evidently "have been formed through metasomatic replacement of the sedimentary material by substance contained in liquid or gaseous solutions."[15] Of these minerals tourmaline, biotite, chlorite, scapolite, garnet, magnetite, amphibole, and pyroxene occur locally.

COPPER.

Disseminated copper deposits, described more fully on pages 115–118, occur in the vicinity of the Snowstorm and National mines and, according to Huston,[16] extend thence northwestward for about 7 miles in certain beds of the formation involved in the Granite Peak syncline. The same writer reports their presence in the Newland, St. Regis, and Revett formations, although only in the Revett have commercial deposits been opened. The two principal deposits are in the Snowstorm and National mines, situated about 1½ miles apart.

The Snowstorm deposit, unique as a large producer of copper ore in a region characterized by large deposits of lead, silver, and zinc, was first exploited in 1903 and ceased to produce in 1915. The mine was opened to a depth of 1,600 feet by four tunnels. It develops a shoot of disseminated copper minerals in a steeply inclined stratum of medium-grained hard white quartzite of the Revett formation. The northward continuation of this particular stratum is not definitely traceable, but it may possibly be opened by the Snowshoe, Missoula Copper, and National workings, and if so has a length of about 1½ miles. The southern limit of the copper-bearing stratum seems to be the Snowstorm fault, which, as forecast by Ransome,[17] cuts off the ore body.

The fault is more properly a fault zone in which the north fault plane is the largest and by bringing different beds of the Revett into contact cuts off the ore zone. Other faults in the zone bring St. Regis beds into contact with Revett beds, the relation implying a reverse movement. Ore has been extracted along a bed 40 feet wide for a distance of 700 feet and down nearly to the 900-foot level. The pitch of the ore shoot is nearly vertical near the surface, but a short distance below it changes to a low eastward pitch. As the fault and copper-bearing stratum converge eastward, it thus happens that a section of nearly barren quartzite lies beneath the ore shoot in the western part of the mine and separates it from the fault.

The principal unoxidized ore consists of quartzite impregnated with minute particles of bornite, chalcocite, chalcopyrite, and tetra-

[15] Calkins, F. C., U. S. Geol. Survey Prof. Paper 62, p. 32, 1908.
[16] Huston, George, The copper beds of the Coeur d'Alene: Min. and Sci. Press, vol. 110, pp. 145–147, 1915.
[17] Op. cit., p. 152.

hedrite. The richest ore is of a uniform dark-gray color; the leaner ore is lighter and finely specked. Oxidized ore largely predominated down to the 600-foot level, although in some places secondary minerals occur down to a depth of 1,600 feet. Malachite is the chief product of oxidation; cuprite occurs in small amounts. The finely disseminated minerals replace the siliceous cementing substance of the quartzite, sericite, and siderite, and to a much less extent the grains of quartz and feldspar.

The structural relations of the Snowstorm deposit are discussed at considerable length in the report by Calkins and Jones,[18] based on field work in 1912. As a result of their study they believe that it "seems reasonable to hope that a continuation of the pay shoot may be found below the No. 4 tunnel level."

The deposit of the National mine occupies a gritty quartzite bed in the Revett formation from 40 to 50 feet thick. It strikes northwest and dips about 45° SW. The tunnel penetrates the Newland, St. Regis, and Revett formations, and the ore-bearing stratum is a part of the Revett. It crosses several faults, the largest of which strikes east-west and dips south. This fault separates the St. Regis from the Revett and probably limits the ore body, though at a considerably greater depth than now attained.

The ore minerals are distributed irregularly through the quartzite stratum but so far as known do not occur north of a quartz vein which has a general northwesterly course and a nearly vertical dip. This vein ranges from a few inches to 4 feet in width and contains scattered chalcopyrite, pyrite, and a little galena. The shoot of disseminated ore is about 250 feet long as worked in 1916, although on the ends ore grades into sparsely mineralized quartzite through so wide a zone that the length of stopes is largely controlled by the price of copper. Most of the ore is fine-grained gritty gray quartzite minutely specked with chalcocite and chalcopyrite. The sulphides occur for the most part interstitially, but they also replace the quartz and feldspar grains. Both the disseminated chalcocite and chalcopyrite appear to be primary minerals, as chalcopyrite characteristically forms an envelope around chalcocite and no instance was observed of chalcopyrite partly replaced by chalcocite. In a crushed zone along the footwall chalcocite occurs in seams and minute veinlets cutting disseminated ore, and this chalcocite is clearly a product of secondary concentration. In polished section it may be seen to replace chalcopyrite. Throughout the mine the ore is of much lower grade than the Snowstorm ore, and only under particularly favorable conditions has profitable extraction been possible.

Other occurrences of disseminated copper minerals have been disclosed by the Snowshoe and Missoula Copper prospects. In the

[18] U. S. Geol. Survey Bull. 540, pp. 205–207, 1914.

Snowshoe the copper-bearing stratum is about 30 feet thick and contains sparsely and irregularly disseminated copper minerals at two horizons in the Revett quartzite. The abundance of mineralization in the principal bed is closely related to its abundance in a quartz vein which traverses the bed. In places where the vein contains much iron and copper sulphide the same minerals occur in the adjacent quartzite in diminishing amounts to a distance of 30 feet or more.

SPHALERITE AND GALENA.

Minute grains and clusters of galena and sphalerite occur adjacent to most of the veins and it is probable that systematic microscopic examination of the wall rock throughout the area would show these minerals, particularly sphalerite, to be much more widely disseminated than is now known. The most extensive known deposit of them is in the east end of the Bunker Hill and Sullivan mine, as described in considerable detail by Hershey.[19] Here sphalerite is much more abundant than galena and is scattered through steeply dipping beds of the Revett quartzite from 5 to 30 feet thick and in ill-defined patches irregularly spaced. The best exposures are in the middle Blue Bird tunnel. Here also are numerous narrow veins of pyrite, containing scattered bunches of brilliant galena and a few crystals of pyrite and patches of sphalerite. Thin sections show all gradations between the sulphides in the veins and those in the wall rock, but in the wall rock sphalerite is locally most abundant, whereas in the veins pyrite predominates. Sericite and sparse siderite occur in both and were in part formed contemporaneously with the sulphides. Most of the disseminated sphalerite is interstitial with respect to the grains of quartz, but some of it clearly replaces quartz, projecting as irregular embayments into rounded grains of the quartz. Galena envelopes the zinc sulphide and rarely may be seen to traverse grains of it. Chlorite, in no place abundant, is intergrown with sericite in scattered nests and clusters of crystals. Tourmaline has been reported,[20] but none appears in material examined by the writers.

Specimens containing disseminated sphalerite, generally subordinate to pyrite, were collected in the Ontario, Senator Stewart, Hecla, Morning, Helena-Frisco, and Success mines. In some of them galena also is present, and in those from the Helena-Frisco and Success mines magnetite occurs in disseminated form. Magnetite, chalcopyrite, and galena occur in a specimen from the Carbonate Hill mine. The disseminated minerals seem to be most common in the more gritty quartzite beds, but in the Helena-Frisco and Success mines

[19] Hershey, O. H., Genesis of lead-silver ores in Wardner district, Idaho: Min. and Sci. Press, vol. 104, pp. 750–753, 786–790, 825–827, June 1, 8, 15, 1912.
[20] Hershey, O. H., op. cit., p. 752.

they occur with much pyrite in sericitic slaty material and in the granitic rocks. Many of the thin sections from these mines reveal minute quartz veinlets adjacent to which the disseminated specks of sulphide are more closely spaced.

PYRITE.

Pyrite is very common along the margins of ore bodies throughout the county, grading irregularly outward from a massive form that is clearly a part of the lode into disseminated grains which vanish within distances of 25 feet or less. In the Bunker Hill and Sullivan mine, however, it is much more widespread, as is indicated by Plate VIII (p. 52). In most places where pyrite is abundant in the lodes it is also disseminated in the wall rock. In the Bunker Hill and Sullivan mine it attains greatest prominence in the general area of the Blue Bird seams, though by no means confined to their immediate vicinity. Here, however, the beds are steeply inclined, and solutions following the narrow Blue Bird fractures had abundant opportunity to depart from them and work upward along the beds of quartzite. If the disseminated pyrite worked outward from these seams during their formation, as is believed, it is noteworthy that the sphalerite and galena which they contain did so locally. This, however, is not an unusual phenomenon, for in many gold deposits pyrite has been deposited much farther from the fissures than any other of the ore minerals.

Some thin sections of the ore minerals show that siderite was formed contemporaneously with pyrite, although much of it may be a little older, as shown where the two minerals form intersecting veinlets.

The disseminated pyrite occurs in minute cubes, many of them perfectly formed, which stud the sericitic matrix of the quartzite and, though seldom distinguished megascopically, impart to it a dark-gray color. Many of the cubes transgress the boundaries of quartz grains and locally exceed the largest quartz crystals in size, clearly showing that they were formed by metasomatic replacement.

In the drainage basin of St. Joe River pyrite occurs disseminated in many localities, not only in the Burke and Revett formations and the underlying quartzite and schist but also in the Newland beds.

SIDERITE.

Siderite, the characteristic gangue mineral of the Coeur d'Alene lodes, is widely distributed in disseminated form. In discussing its occurrence Ransome [21] says:

[21] Op. cit., p. 97.

The mineral, though especially abundant near certain areas of intense mineralization, is by no means confined to the ore bodies or to their immediate vicinity but is widely distributed through the quartzite rocks in the district. Mr. Calkins finds, however, that although siderite has a much wider distribution than the ores, yet it is most abundant where the rocks have been most folded and fissured.

In the Canyon Creek area siderite is abundant in the Burke and Prichard formations; near Wardner in these formations and also in the Revett and St. Regis; and in the upper St. Joe River basin siderite is "generally disseminated in the sandstone of the strata representing the lower part of the Newland ('Wallace') formation."[22]

The mineral is clearly due to replacement of the inclosing rock as determined by Calkins,[23] who says:

Siderite occurs in individuals of varying size; but probably the majority of them are larger than the clastic grains of the rock, the mineral being in places crystallographically continuous for a diameter of several millimeters. The form of the individuals varies much in regularity. The outlines of many of them, especially of the larger grains, appear ragged, but in such cases the boundaries are seen on close examination to be determined in large part by rhombohedral planes. In other individuals the rhombohedral form of the crystals is more marked, in some being perfect. It is an important fact that in any specimen the crystallographic boundaries can be seen cutting sharply across quartz grains, which proves that the siderite has been developed by molecular replacement of the quartz and not by filling in of cavities. The larger grains in the sandstones and quartzites frequently inclose grains of quartz and, much less commonly, scales of sericite. It is thus evident that the siderite replaces the fine-grained cementing material more readily than the larger grains of quartz.

The disseminated siderite is seldom preserved in rocks at or near the surface. The mineral readily breaks down under the influence of weathering into limonite, which in larger part, as rusty specks or patches, occupies the position of the original siderite grains and gives a most characteristic mottled appearance.

Contact-Metamorphic Replacement Deposits.

Contact-metamorphic replacement deposits are not well developed in Shoshone County, although a number of the lodes fulfill part of the criteria of contact metamorphism and evidently were formed under comparable conditions of temperature and pressure. It is probable that had the mineralizing solutions invaded a series of limestones instead of quartzites and siliceous slates this group would have been much more fully represented. In a limestone area much of the material entering into the lime silicates is present in the invaded rocks, but in quartzite the lime, in addition to the silica and iron, which

[22] Pardee, J. T., Geology and mineralization of the upper St. Joe River basin, Idaho: U. S. Geol. Survey Bull. 470, p. 43, 1910.
[23] U. S. Geol. Survey Prof. Paper 62, p. 97, 1908.

are commonly in excess in contact zones even in limestone, must also have been supplied by the magma. Furthermore, quartzite is less susceptible of irregular replacement than limestone.

In view of these considerations the contact deposits are of particular importance in a systematic study of the mineralization of the county, even though they have contributed comparatively little to its total production.

The Success mine, formerly the Granite, is an important producer of lead and zinc ores and the principal representative of the group of contact deposits. The Gem and Frisco lodes locally contain lime silicates in intimate association with metallic sulphides and magnetite and are considered to be closely related to typical contact deposits in origin. Similar relations exist in the Rex and Sunset deposits on Ninemile Creek and probably also in the Copper Prince deposit in the "southern copper area."

GEOLOGIC RELATIONS.

The deposits occur in Prichard slate, in Burke quartzite, and in the calcareous shales of the Newland formation at or near their contact with granitic rocks.

The Success mine, on the East Fork of Ninemile Creek about 5 miles north of Wallace, is in the central part of a narrow tongue of Prichard slate and quartzite which extends eastward for nearly a mile into the monzonite mass west of Gem. In the vicinity of the mine the sedimentary beds are metamorphosed to biotite schist, micaceous quartzite, and dense greenish quartzite, and each type contains variable amounts of garnet, pyroxene, biotite, muscovite, and locally chlorite and epidote. The structure of the beds is in many places difficult to decipher, but in general they strike northwest and dip 30°–85° SW. The Prichard rocks are intricately traversed on nearly all the thirteen levels of the mine by tongues and dikes of the monzonite, as is well illustrated on the 300-foot level. (See fig. 2, p. 26.) In places the monzonite apophyses are traversed by the ore, as is clearly shown between stations 300 and 308 on this level, where a 10-foot dike nearly at right angles to the course of the ore body is exposed in opposite sides of the stope at a point where ore was extracted to a width of about 12 feet. Another equally clear example of ore formed subsequently to the monzonite appears on the same level near station 324, where an offshoot from the main ore body replaces a 4-foot dike and extends into the quartzite beyond.

On the 100-foot level, 191 feet above the 300-foot, the ore body, retaining its width of 3½ to 5 feet, crosses a southwestward-dipping monzonite dike. This relation continues for 60 feet below the level, but thence downward no ore has been found in the monzonite, although ore has been mined in the adjacent quartzite for more than 100 feet deeper. Above the level the stope continues between granite walls for 40 feet without diminution in width. As there is notably less fracturing within the dike than beyond it, the history is interpreted as (1) shearing, (2) intrusion of the dike, and (3) renewed shearing accompanied or followed by replacement of the wall rock, both quartzite and monzonite, by ore.

The relations below the 100-foot level described above suggest that the quartzite was more easily replaced by the mineralizing solutions than the monzonite, which is clearly shown by hand specimens to have been locally replaced. This suggestion is borne out by the deposit as a whole, for, although the monzonite has been intensely altered in places, the total volume of it which has been replaced by ore, compared with the other rocks, is negligible. Of the several types of rock in the mine, most of the ore is inclosed in mica schist, a notable amount in the micaceous quartzite, very little in the dense greenish quartzite, and still less in the monzonite.

The Sunset mine, near the summit of Sunset Peak, develops a lode from 2 to 10 feet wide which extends from the monzonite westward almost at right angles to its contact, across a patch of eastward-dipping Burke beds into the underlying Prichard formation. All the known ore is inclosed in the sedimentary beds, although on an adjoining claim to the east a small tunnel reveals seams of galena and pyrite in the monzonite. The main tunnel follows the monzonite contact for several hundred feet but diverges from it before reaching the lode, and in no place along the lode has the contact been explored. Garnet, a greenish pyroxene, and epidote are present in much of the material on the dump, in some specimens in intimate association with sulphides. The granitic rock, which here has the composition of syenite, contains much secondary epidote and pyrite.

The Frisco lode extends from the irregular monzonite mass on a general course almost at right angles to its border. The deposit (described on pp. 100–102) is not typical of contact metamorphism, yet in parts of the mine actinolite, epidote, and magnetite are so intimately associated with the ore that it is believed to have formed under conditions closely approaching such phenomena. It is noteworthy, on the one hand, that a lamprophyre dike, considered to be a basic differentiate from the monzonite, cuts cleanly across the lode, and on the other that the monzonite is locally replaced by epidote, pyrite, and sphalerite.

The Gem lode is probably a faulted segment of the Frisco and bears similar relations to the monzonite.

The Rex lode, situated in the same area of the Prichard formation as the Success, is not in immediate contact with the monzonite at

any place in the mine as now developed, although a small stope on
the North vein, 200 feet from the Rex lode, has hornblende monzonite on the hanging-wall side. Both lodes extend westward almost
at right angles to the general course of the monzonite contact. Metamorphism is recorded by a local intimate association of epidote.
garnet, and blue amphibole with the sulphides.

The Copper Prince deposit is in the greatly metamorphosed Newland formation, which occurs widely in the vicinity of the granitic
masses of the southwestern part of the county. The lode, which is
traceable for about 2,000 feet, is inclosed in the sedimentary rock in
most places but locally occupies a shear zone along the igneous contact. The deposit was not visited during the recent examination.
and available information does not include data on the detailed relationship of sulphides and lime silicates. Its assignment to this
group is provisional.[24]

DISTRIBUTION AND FORM.

In each of the contact-metamorphic and closely related deposits
the position of the ore bodies is controlled by a zone of shearing or
crushing, which, except at the Copper Prince, extends almost at right
angles to the igneous contact. Even though such zones of fracturing have served to direct the ore-forming solutions, the size of the
ore bodies is due largely to replacement of the inclosing rock. The
ore bodies are in general irregularly lenticular, and in most of them
the vertical axis is much greater than the horizontal. In many places
the ore shoots branch in the plane of the shear zone, giving more
than one stope on certain levels. In the Rex lode, for example, there
are two prongs to one general ore shoot. The eastern one has been
stoped down to No. 2 level; the western one down to the 100-foot
level and in part between the 100 and 300-foot levels. On the eastern
prong a little stoping has been done from the 300-foot level also, but
below, on No. 5 and No. 6 levels, development indicates that the two
shoots have merged. A similar irregularity is shown in the outline
and distribution of ore bodies in the Helena-Frisco mine and in the
Copper Prince deposit.[25]

The form and distribution of the ore bodies of this group of deposits are well illustrated in the Success mine. Here the distribution
of the ore bodies is controlled by the shearing planes of biotite schist,
although the ore is by no means confined to such planes. This member of the Prichard, from 10 to 20 feet thick, strikes a little west
of north in the southeastern part of the mine, but farther north

[24] Pardee, J. T., Geology and mineralization of the upper St. Joe River basin, Idaho :
U. S. Geol. Survey Bull. 470, pp. 49–50, 1911.
[25] Idem.

turns abruptly westward and terminates against the monzonite in the vicinity of the shaft.. Here it dips about 80° S., but toward the southeast it gradually flattens to a dip as low as 45°.

The ore occurs in lenses in the biotite schist and in the micaceous quartzite above and below it. The lenses attain their maximum development near the bend in the formation, and here an outer and an inner series occur—the former principally in the schist and the latter in the overlying micaceous quartzite. Toward the west the ore extends into the monzonite for a short distance in places, but toward the southeast the mineralization dies out through a series of isolated lenses on the same general zone. Most of the lenses in the productive part of the zone are connected either by merging in places along their periphery or by oblique or transverse bodies. Thus if a stope sheet were made by projecting the worked parts of the deposits to a common plane, the ore body would appear as a vein. On the other hand, cross sections and plans of separate levels bring out the essential characteristics of the irregular replacement phenomena controlled by the distribution of a zone of shearing in the comparatively thin biotite schist member of the Prichard formation.

The principal ore body was found on the curved part of the zone and was worked continuously from the surface to the 1,200-foot level. It was largest on the 500-foot level, where the sill floor was 60 by 100 feet; above and below this level it ranged from 125 to 235 feet in length and from 16 to 18 feet in width.

Within the lenses there are many residual masses of country rock, locally giving faces the appearance of a sulphide-cemented breccia. Elsewhere great isolated blocks or " horses " occur in the ore bodies. In most places these are mined as a necessity of the shrinkage system employed, but locally they stand in stopes now worked out. Evidence of fissuring along the zone of shearing is entirely lacking, although it seems likely that some relative displacement of the quartzites above and below the schist accompanied the development of schistosity. The importance of replacement in the creation of the ore bodies is shown by innumerable expansions of ore directly across the schistosity with veinlets extending out along the cleavage planes. Wider parts of the ore bodies in many places extend beyond the limits of the biotite schist into the micaceous quartzite above and below, and locally lenses of considerable size occur many feet beyond the limits of the schist.

The Sunset, Rex, and Helena-Frisco deposits are entirely similar to the Success except that they cut across the inclosing beds and their distribution seems to be more nearly independent of the composition of the wall rock.

THE ORES.

The ore from the contact and related deposits is valuable principally for zinc and lead. The Success, Rex, and Helena-Frisco were formerly worked for lead but have yielded an increasing amount of zinc in depth and should now be considered zinc mines. The Copper Prince is a copper deposit.

The ore of the Success mine, which was studied in greater detail than any other of the group, may be considered as representative of all the deposits except the Copper Prince. As extracted by a shrinkage system without sorting, it contains from 8 to 10 per cent of zinc, 2 per cent of lead, and about $1\frac{1}{2}$ ounces of silver to the ton. Sphalerite is the dominant mineral, but magnetite, pyrite, galena, and quartz are present in most places. The characteristic gangue is quartzite or biotite schist replaced to a greater or less extent by lime silicate minerals. The sulphides are distributed throughout the ore body in blebs, bunches, lenses, and connected areas of ramifying habit. The galena was clearly formed later than the sphalerite, which it traverses as veinlets and connected patches. Pyrite is scattered through the sphalerite in isolated crystals or small groups of crystals and may be contemporaneous with it. It is noteworthy, however, that the pyrite extends into the surrounding rock much farther than the sphalerite, in the monzonite replacing the hornblende to a distance of at least 200 or 300 feet from any known ore body. Magnetite is intimately associated with the sphalerite and may be nearly contemporaneous with it, although in one of the specimens studied the magnetite is distinctly the older.

Sericite is associated with the sulphides irrespective of whether they replace quartzite, schist, or monzonite. In the quartzite it occurs as veinlets cutting the quartz grains and interstitially between them; in the schist as flakes parallel to the schistosity; and in the monzonite as isolated flakes and felted aggregates in feldspar areas and as veinlets. A pleochroic carbonate with the low index higher than Canada balsam but much lower than siderite, and hence probably a ferruginous calcite, accompanies the sericite in one of the specimens. Where a veinlet of calcite intersects one of sericite the latter invariably cuts the former, and where the two minerals occur in the same veinlet the sericite occupies the medial plane. Quartz occurs both in veinlets cutting the ore and in microscopic intergrowth with the several sulphides.

Silicate minerals occur in much of the ore in microcrystalline individuals and aggregates closely intergrown with the sulphide minerals. In the monzonite garnet, pyroxene (probably hedenbergite), a blue amphibole, epidote, chlorite, and rarely tourmaline occur in minute microscopic grains. Garnet is the prevalent contact mineral in the quartzite and schist, but epidote, diopside, and chlorite are present in most places in variable amounts. An intimate intergrowth of the silicate minerals and sulphides is shown in many parts of the deposit.

METAMORPHIC PHENOMENA.

General features.—Unlike the results of metamorphism in limestone areas, only slight changes in general aspect are shown by the quartzite rocks here found. Even where the metamorphism is intensest the texture is little changed, and difference in color, combined with a slightly increased specific gravity, is the principal indication of transformation. Where garnet is abundant the color is either pink or light amber; where epidote or a pyroxene is abundant the color is greenish. Minor amounts of these minerals do not noticeably affect the gray or greenish-gray color of the quartzite, although such material has a flinty aspect due to advanced recrystallization of the siliceous mass and readily suggests the name hornstone, which may be appropriately applied to it.

In the Newland rocks traversed by the Copper Prince deposit the metamorphism has accomplished changes more like those characteristic of limestone areas. Garnet, scapolite, pyroxene, and amphibole are in many places developed to such an extent that the subdivisions of the Newland formation lose their normal lithologic character and mapping of them is extremely difficult.

Locally the monzonite is metamorphosed near its periphery, particularly in the Success and Helena-Frisco mines. It gives rise to a rock somewhat resembling the intensely metamorphosed quartzite but in most places retains sufficient of the monzonite pattern to be easy of identification. Where the transformation is less advanced the rock has a dull appearance due to alteration of the feldspar crystals and is usually specked with pyrite, which replaces hornblende. Thus within some of the deposits two common types of metamorphism occur—exomorphism, or the transformation of the invaded rock, and endomorphism, or the metamorphism of the intrusive rock. These phenomena were observed in all the deposits except the Copper Prince but were studied in detail only in the Success mine. A record of observations made on this deposit may be considered fairly representative of the group.

Exomorphism.—The Prichard rocks throughout the Success mine have been " completely recrystallized to an aggregate of interlocking quartz grains which inclose variable proportions of pale-green monoclinic pyroxene, green-brown biotite, white mica, * * * and garnet." [26] Although the metamorphism is general throughout the tongue of slate and quartzite which extends into the monzonite,

_____

[26] Ransome, F. L., op. cit., p. 185.

different beds have been affected differently. Garnet is most abundant in the fine-grained greenish quartzite, which occurs principally in the eastern part of the mine. Sericite is conspicuously developed in the micaceous quartzite above and below the schist and to a less extent in the other rocks. Diopside occurs principally in the biotite schist. Tourmaline appears in one of the slides of greenish quartzite. The biotite in the schist and the muscovite in the adjoining quartzite are oriented parallel to the foliation and probably were formed in principal part incidentally to the shearing stresses that caused the development of schistosity. The lack of shearing within the monzonite where it truncates these beds leads to the conclusion that the development of the schistose zone and of the biotite and muscovite incident thereto preceded the intrusion. The development of the sericite, garnet, pyroxene, and epidote, however, was a distinctly later phenomenon. These minerals occur in irregular bunches, veinlets, and ill-defined patches connected by stringers of similar material. Their general distribution is controlled by the zone of shearing, but in many places they lie athwart the foliation, whose planes terminate abruptly against them as against the sulphide bodies.

The areas of moderate metamorphism are more extensive than those of replacement by ore, but areas of intense metamorphism, in which lime silicates constitute 50 per cent or more of the mass, seem to be much smaller than the ore bodies. The rocks are more intensely metamorphosed in the east end of the mine than elsewhere, but most of the ore occurs in the west end.

The essential contemporaneity of the sulphides and silicates is proved by specimens from several parts of the deposit, which abundantly bear out Ransome's conclusion [27] that "the association of the ore minerals with the metamorphic silicates is so close that the conclusion of their contemporaneous genesis is unquestionable." In a few places (see Pl. VI) the sulphides form veinlets in the metamorphosed rock, suggesting that the sulphide deposition continued after the cessation of lime silicate development. The relation, however, is believed to be exceptional, although the amount of material available for microscopic study does not warrant a definite conclusion. The deposit is not favorable for paragenetic studies without the aid of the microscope, as the contact-metamorphic rock in most places is megascopically uncertain of identification and impossible of classification. The metamorphic silicates in most specimens occur in exceptionally minute grains, a magnification of 50 to 250 diameters being necessary to the recognition of separate crystals. Interlocking with these in low-grade ores are equally small grains of sphalerite, pyrite, and less commonly galena.

---

[27] Op. cit., p. 185, pl. 10.

Endomorphism.—The most interesting result of the recent investigation of the deposit is the recognition for the first time of clear evidence of endomorphism. The monzonite in many places in the mine has been changed from a fresh hornblende-bearing rock into an intensely sericitized pyrite-bearing rock and locally into an intimate intergrowth of sulphides and metamorphic silicates. This conclusion is opposed to that of Hershey,[28] who states that most of the sulphide patches in monzonite " have fragments of schist attached, indicating that they are inclusions " and thus are not discordant with his general proposition that " the veins are older than the intrusion of the adjacent monzonite." It agrees in general with the view of Ransome,[29] who states that the ore " was deposited shortly after the intrusion of the monzonite and is a phase of contact metamorphism." His observation, however, based on relations as exposed in 1904, that " the ore is strictly confined to the sedimentary rock," now requires modification.

The diversity of opinion of the previous writers as to the relative age of the ore and the monzonite justifies a rather complete assembling of the evidence for the writers' conclusion that the metamorphism and metallization were subsequent not only to the injection of the magma but to the chilling and solidification of its outer part.

As previously stated and as illustrated in figure 2 (p. 26), it is clear that ore bodies cut apophyses of the monzonite and are of later development. This conclusion is borne out by the evidence of hand specimens, thin sections, and polished surfaces. Plate VII, *A*, is a photograph of a specimen that shows no included sedimentary rock and consists of sphalerite, pyrite, and galena in a gangue of metamorphosed monzonite, in which magnetite is locally abundant. Thin sections prove the essential contemporaneity of the metamorphic silicates and sulphides. In one part of the specimen an aplite dikelet cuts the monzonite and is itself cut by a seam of pyrite, magnetite, sphalerite, and galena. Throughout part of its extent magnetite, sphalerite, and galena are collected along the contact of the dikelet with the inclosing rock. In many parts of the specimen the sulphides have irregular replacement contacts against the monzonite, which is cut locally by minute veinlets of the sulphides. Thin sections were made from the other part of the specimen at points corresponding to those marked A, B, C, and D in the illustration. Thin section A consists of the garnet andradite, a green and blue pleochroic amphibole, hedenbergite, sphalerite, pyrite, magnetite, galena, and locally quartz. The silicates and sulphides occur predominantly in interlocking grains, but in a few fields sphalerite and galena form veinlets. Section B consists of an intergrowth of garnet, epidote,

[28] Op. cit., p. 18.
[29] Op. cit., p. 185.

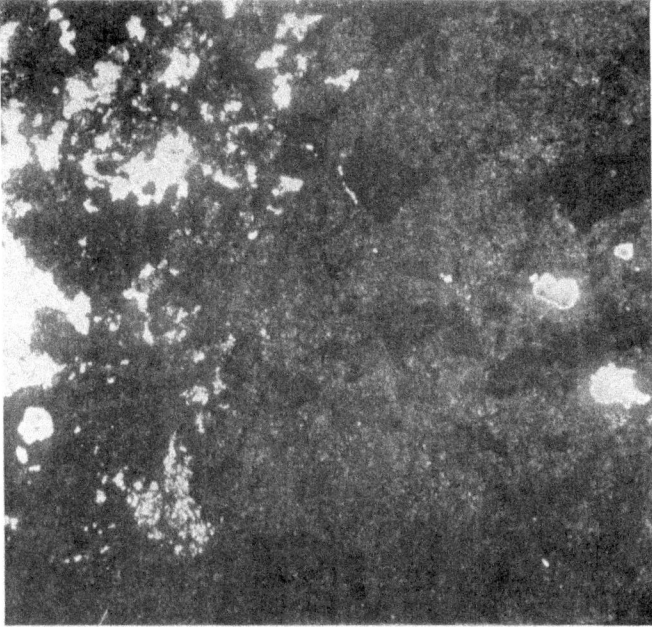

A.

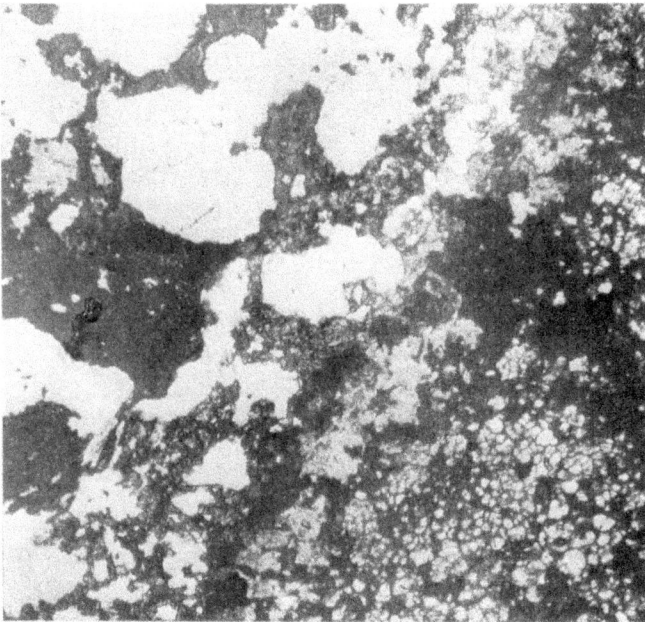

B.

PHOTOMICROGRAPHS OF METAMORPHIC ORE, SUCCESS MINE.

ORE IN MONZONITE AND SYENITE.

A. Monzonite partly replaced by sphalerite (sp), galena (G), pyrite (P), magnetite (part of black area), and lime silicates after injection of an aplite seam shown at B. Thin sections were cut at A, B, C, and D. From point near shaft on 860-foot level, Success mine. $^3/$, natural size.

B. Sphalerite, galena, and pyrite replacing monzonite (light). Success mine, 300-foot level. $^4/_5$ natural size.

C. Galena and chalcopyrite (black areas) replacing syenite. From the Lily prospect. $^4/_5$ natural size.

hedenbergite, green and blue pleochroic amphibole, pyroxene, garnet, and sericite in the feldspar areas. Section C is made up of minute interlocking crystals of garnet, blue-green amphibole,[30] hedenbergite, epidote, sericite, biotite, olivine, and chlorite. The slide from point D consists of much less altered monzonite, in which feldspar forms perhaps 80 per cent of the mass. Sericite is the dominant alteration product and occurs in seams, flecks, and patches in both orthoclase and plagioclase areas. Chlorite occurs in seams and with what appears to be iron oxide in areas corresponding in distribution to the hornblende of the fresh monzonite. Sphalerite occurs in seams and with pyrite along one edge of the section.

The specimen described above was taken in a stope from the 860-foot level, but others equally conclusive of metamorphism and ore deposition after marginal solidification of the monzonite were collected from the 100, 300 (fig. 2), and 1,000 foot levels.

The endomorphism at the Success mine is perhaps more intense than elsewhere in the county, but specimens from the other contact deposits show similar phenomena, and in certain places where hornstone is abundant, although lime silicates were not observed, there is a similar age relation of ore and igneous rock. In the American prospect, immediately west of the Success mine, seams of quartz, galena, and specularite traverse monzonite in which sericite, chlorite, and pyrite are developed. Similar relations occur in the Lily and Nelly Pollar prospects, east of Murray. In the Lily the ore consists of galena with associated chalcopyrite replacing syenite adjacent to a 2-inch clay gouge which separates it from Prichard slate. The bunches and seams of sulphides are most numerous next to the gouge, but they are present in diminishing numbers to a distance of 20 feet from it in two crosscuts 65 feet apart. The Sunset, Rex, and Helena-Frisco deposits are more like the Success, in that lime silicate minerals are locally well developed as alteration products of the monzonite.

Replacement Deposits Along Fissures and Fracture Zones.

Nearly all the ore deposits of present economic importance in Shoshone County were formed by replacement along fissure and fracture zones. Irregular contact deposits, such as occur commonly near igneous contacts in limestone areas, are poorly represented, and fissure fillings, typified by many veins elsewhere in the State, are here but meagerly developed. To this group belong most of the gold deposits near Murray, which have been almost abandoned for many years.

---

[30] The amphibole is optically positive, has positive elongation, extinction to 30°, and is strongly pleochroic in X yellowish to greenish, Y yellowish-blue, and Z blue tones.

The more important deposits of the area, appropriately termed lodes, are great lenses elongated vertically and very irregular in detailed outline. Along many of them stringers and bunches of sulphides extend sporadically into the wall rock well beyond the limits of profitable mining. Locally, as in the Morning and Standard-Mammoth mines, great horses occur completely surrounded by ore, and in almost any stope unreplaced remnants of country rock may be found in the lode. Most of the fracture zones record more than one movement, later minerals occupying seams that traverse minerals formed earlier. It is these recurrent movements during the general period of ore deposition that make the broader facts of paragenesis particularly easy of recognition. None of the movements seem to have caused notable displacement of the strata, although in an area of formations of great thickness, made up of many beds almost identical in appearance, an offset of 100 feet or more may not be detected. Movements during deposition, however, were certainly small, as they resulted in a shattering of the older minerals without developing a gouge in them.

LEAD-SILVER DEPOSITS.

IMPORTANCE.

The lead-silver deposits of Shoshone County have contributed predominantly to its total production, to the end of 1920, of $487,299,838 worth of metals. Of this amount more than half has come from lead, and about one-fourth from silver, principally associated with the lead. Although gold was discovered first and was the only metal produced during 1884 and 1885, lead and silver became noteworthy factors in 1886 and with only a few periods of reversal have steadily increased in output. In 1904 copper became important in the county's production, but it has greatly decreased since the closing of the Snowstorm mine. In 1905 zinc entered the list of metals produced, and it is this metal, if any, which may finally eclipse lead as a resource of the region.

There is every reason to believe that a large production of lead will continue for many years. The Bunker Hill and Sullivan, Morning, Hecla, and Hercules mines all have large ore reserves and probable reserves, and increased output from them has largely offset a decrease due to the failure or diminished production in recent years of the Tiger-Poorman, Standard-Mammoth, Last Chance, Helena-Frisco, and other smaller mines. New discoveries will probably further prolong the period of lead mining in the county, for in an area where such deposits as the Senator Stewart, Caledonia, and Tamarack and Custer lay hidden during 20 years or more of intense mining activity, other notable discoveries must be expected.

DISTRIBUTION.

The most valuable lead-silver deposits in Shoshone County occur near Wardner, where the Bunker Hill and Sullivan, Senator Stewart, Caledonia, Sierra Nevada, and Last Chance are the principal mines, and in the area around Burke, which includes the Greenhill-Cleveland, Standard-Mammoth, Helena-Frisco, Hecla, Tiger-Poorman, Marsh, and Hercules deposits. A belt extending eastward from the Burke area includes the Morning, Alice, and Gold Hunter as the principal representatives. On Ninemile Creek and the head of Carbon Creek are the Tamarack and Custer, Idora and Tuscumbia, Sunset, and California. Outlying deposits that have afforded some production are widely scattered and include the Hypotheek, west of Pine Creek; the Jack Waite, near the head of Eagle Creek north of Murray; the Monarch, Bear Top, and Paragon, near the head of Prichard Creek; the Carbonate Hill, on Willow Creek; and several minor deposits south and west of Wallace. In the Pine Creek district lead is prominent in several of the zinc prospects, and on Slate Creek, a tributary of St. Joe River, many of the prospects contain galena.

RELATION TO ROCKS OF DIFFERENT CHARACTER.

The lead-silver mines are predominantly in the Burke and Revett formations, but there are numerous exceptions. The Gold Hunter is in the Newland, the Carbonate Hill in the St. Regis, the Hypotheek in the middle Prichard, the Monarch, Jack Waite, Paragon, and Murray Hill in the upper Prichard, and the Evolution in the lower Prichard. No lead mines have been opened in the monzonite, but in several places, notably at the Success mine and the Lily prospect, seams of lead ore traverse it. Besides the principal deposits there are numerous others in each of the groups. There is no persistent difference in the character of the ore in any way related to the inclosing formation, and it seems probable that the preponderance of lead mines in the Revett and Burke is due primarily to the occurrence of these formations in the two areas that contain loci of lead mineralization. The absence of lead mines in most areas where these formations crop out shows that the local concentration of immense deposits in them is not a consequence of the character of the rocks themselves. Neither is the localization due to folding or faulting. The area near Burke is not greatly faulted, and the folds are more open there than in many other parts of the region. On the other hand, the Wardner area is closely folded and broken by faults in a manner comparable to the partings in schist. Areas of nonproductive Burke and Revett rocks are known which represent all gradations between these structural extremes. Surely this means that the fundamental cause of the localization of the ore bodies is neither the character of the inclosing rock nor the extent of fracturing, although both unquestionably exert a modifying influence. These features are further discussed in the section on ore genesis (pp. 141–150).

The distribution of ore bodies is in part related to rock texture, as pointed out by Ransome,[31] who observed that

> The rock most favorable to the formation of large bodies of lead-silver ore is undoubtedly a fine-grained sericitic quartzite in which the quartz grains instead of closely interlocking are separated * * * by considerable interstitial material, largely sericite. A large part of the Revett quartzite, however, is not particularly sericitic and has grains that interlock after the manner of typical quartzite. Such rock seems to be less favorable to the deposition of lead-silver ore than the more sericitic variety, which, while not perhaps the most characteristic part of the formation, is very abundant, particularly in the lower beds. Even in those deposits which occur in the Prichard formation, it is noteworthy that the quartzitic beds are the ore bearers and that nearly all the prospects are in the upper part of the Prichard, in beds that partake of the character of the overlying Burke formation.

Although the general truth of these observations has been borne out by the active exploitation of the decade since they were made, several noteworthy exceptions are now known. Unquestionably solutions following any fissure that traverses rocks of different character will replace certain beds more readily than others, but some of the great ore bodies recently opened, notably the Interstate-Callahan, are in dense quartzitic shale, the particles of which are minute and closely assembled. Also several of the mines on Pine Creek contain deposits which replace compact slates. These facts lead to the broad conclusion that given a fissure or fracture zone followed by ore solutions, any rock in the region may contain commercial deposits. Up to the present time no large lead-silver mine has been opened in Prichard slate, but the writers can discover no reason which appeals to them as valid why such deposits may not be found. Neither has lead-silver ore been found in commercial quantities in the granitic rocks; for this the explanation may be as follows: It is shown on page 25 that veinlets of galena and other sulphides traverse the monzonite, but it does not follow that the monzonite had entirely solidified at the time the ores were deposited. How far solidification had progressed is not known, and hence it can not be foretold to what distance from its margin ore may be found in the monzonite masses. In at least two places, however, lamprophyre dikes, doubtless late differentiates from the magma, traverse ore bodies, thus showing that magmatic activity continued at least locally after lead-zinc deposits of large size had been completed.

---

[31] Ransome, F. L., op. cit., pp. 106–107.

CHARACTERISTIC FEATURES.

The lead deposits characteristically replace quartzite and consist dominantly of galena in a siderite gangue with subordinate quartz. Pyrite and sphalerite are present in variable amounts, pyrrhotite and chalcopyrite occur locally, and in a few lodes barite is conspicuously developed.

The deposits are tabular, the replacement having progressed from fissures, as in the Morning, Standard-Mammoth, and other mines; from partings in a sheeted zone, as in the Hecla and Gold Hunter; or from fractures in a crushed zone, as in the March shoot of the Bunker Hill and Sullivan mine. In some of the lodes a banded arrangement of the minerals indicates recurrent opening of the fractures, a feature typified by the Blue Bird seams of the Wardner area.

The prevailing strike of the lead deposits is about N. 70° W., with steep dip to the southwest. There are, however, many exceptions to this attitude. The Alice, Lily, Senator Stewart, Sierra Nevada, East Caledonia, and Jersey fissures of the Bunker Hill and Sullivan mines strike northeast. The principal Bunker Hill and Sullivan ore body strikes N. 42°–45° W. With the exception of this ore body, which dips 38° SW., and the Sierra Nevada, which is locally almost flat, the lodes are nearly vertical. Variations from the vertical in very few of the fissures exceed 20°, and in most of the principal mines the dip is southward.

The ore bodies are distributed along fissures that were doubtless formed incidentally to the faulting of the region, but none of them show evidence of marked displacement, as pointed out by Ransome.[32] The region strikingly illustrates the fact that great faults are not generally mineralized, a fact recognized as due to the heavy gouge accompanying them. The ore shoots, many of them exceptionally large, occur without any recognizable order of distribution along the fissures. They vary widely in form and in size. In the Canyon Creek area the ore shoots are tabular and wedge out on the ends into fissures that are in many places almost devoid of ore minerals, but in the Last Chance and parts of the Bunker Hill and Sullivan mines the ore grades out into country rock both along the walls and at the ends of the shoots. The March shoot of the Bunker Hill and Sullivan mine is unique among the ore bodies of the county. It follows a triangular block of shattered quartzite bounded by faults known locally as the Sullivan, Flint, and Upper Cate. The shoot has been followed from the surface to the fourteenth or lowest level in the mine, a distance on the pitch of the shoot of about 4,000 feet, and throughout it holds closely to the same block of quartzite, nowhere

---

[32] Op. cit., p. 134 and elsewhere.

passing below the Sullivan nor above the Upper Cate faults but in places extending from its normal position in the hanging wall of the Flint fault to positions well in its footwall.

MINERALOGY.

The lead-silver ores consist predominantly of argentiferous galena in a gangue of siderite or quartz, or both. Associated with the galena are sphalerite, pyrite, pyrrhotite, chalcopyrite, stibnite, tetrahedrite, and magnetite. In a few of the deposits barite is conspicuously developed, and locally calcite is abundant. The siderite throughout the district is manganiferous. Silver accompanies the galena in all the deposits, and the tetrahedrite in most places is rich in silver. In concentrating the ores more zinc than lead is lost, and in such tailings the ratio of silver to lead is higher than in the original ores, suggesting that some silver occurs in the sphalerite. This may not be true, however, because in wet concentration tetrahedrite slimes much more easily than galena and a greater percentage of it goes into the tailings.

The several minerals occur in nearly all proportions in different deposits. The Hunter lode and the east end of the Morning lode are particularly high in barite. Tetrahedrite, widely distributed in the lead-silver deposits, is particularly abundant in the Caledonia and Hypotheek and the deposits between Wallace and Wardner. Chalcopyrite is present in most of the veins in small amounts and is a noteworthy mineral of the ores from the Hypotheek mine, the Lily prospect, the Black Bear Fraction, and parts of the Caledonia. Pyrite is nearly everywhere present but occurs in very small amounts in the March shoot of the Bunker Hill and Sullivan mine, in the Jack Waite, and in much of the Standard-Mammoth lode. Sphalerite also is universally present in the lead-silver ores, but in amounts which differ widely in different lodes and in different parts of the same lode. An increase in the amount of zinc with increased depth is generally recognized by the mine operators of the area, although detailed data are not available. During the years 1915 and 1916 the high price of zinc encouraged its recovery, and the production for these years affords an idea of its relative amount in some of the deposits. None of the principal mines near Wardner contain enough zinc to warrant its recovery, although in most of the lodes sphalerite can be recognized. In the Canyon Creek mines and those near Mullan zinc is more abundant. During 1915 and 1916 the Morning and the Greenhill-Cleveland each produced about 1 ton of zinc to 7 tons of lead; in the Helena-Frisco the ratio was nearly 2 of zinc to 1 of lead. The Gold Hunter contains a much lower proportion of zinc than the Morning, and on the other hand the Success contains a higher proportion than the Helena-Frisco. From this the step to typical zinc mines is short, the Interstate-Callahan containing between 4 and 5 per cent of zinc to 1 per cent of lead. It is noteworthy that there is a complete gradation from deposits carrying galena almost to the exclusion of other sulphides, such as the March shoot of the Bunker Hill and Sullivan, through lead-zinc and zinc-lead deposits to rich zinc deposits in which lead is of minor consequence. But more noteworthy still is the occurrence of such a gradation within the same lode. The Morning mine contains much more zinc in proportion to lead in the west end than in the east end, a feature readily observable in the lower levels in 1916. In many of the west stopes the galena occurs as seams and veinlets in a sphalerite mass, whereas the east stopes in many places show sphalerite only as kernels and patches in the galena. This is the more significant because the deposit is entirely within the Revett formation, whereas the other deposits in the Revett are low in zinc. Except for this lode, one of the most productive in the county, and the Hypotheek and Jack Waite, lead mines in the Prichard formation, the deposits may be grouped roughly as zinc in the Prichard, zinc and lead in the lower Burke, and lead in the upper Burke and Revett. With such exceptions—the Morning is an ore body about 1,500 feet long and has been explored to a vertical depth of 3,400 feet without sign of impoverishment—significance can not be placed upon stratigraphic distribution of lead and zinc. There may be some relation between the character of beds and the type of ore deposited in them, but that there is any fundamental dependence of one upon the other is disproved by the Morning, Hypotheek, and Jack Waite lodes.

PARAGENESIS OF THE PRIMARY ORE MINERALS.

The paragenesis of the minerals of the lead-silver deposits is essentially the same whether the dominant gangue, which determines the subtypes, is siderite, quartz, or barite. The prevailing order of deposition of the principal minerals as seen in the stopes and confirmed by a study of polished surfaces was siderite, quartz, sphalerite, galena. The relative abundance of these four minerals and barite gives rise to the different types of lead and zinc deposits. In the March shoot the siderite and galena stages were dominant; in the Hecla the galena and quartz; in the Jersey fissures the quartz and galena, with an earlier siderite filling largely obliterated by the quartz; in the Gold Hunter the siderite, barite, quartz, and galena stages were dominant; in the Greenhill-Cleveland the quartz, galena, and sphalerite; and in the west end of the Morning the siderite, quartz, sphalerite, and galena. From this the transition is short to such a typical zinc lode as the Interstate-Callahan, in which the sphalerite stage is dominant but the other three are well represented. Each of the four minerals is present in all the deposits, except that sphalerite may be absent in the Hypotheek lode, and in dominant amount each has the same age relations. Recurrence of conditions favorable to the deposition of each of the minerals, particularly of siderite and quartz, occurred in several of the lodes, but the relative volume of the particular mineral deposited as a second generation is nowhere large. Thus in the Standard-Mammoth, Morning, and Hecla lodes seams of quartz cross ore in which galena has a mesh-like distribution in older quartz and siderite. Where sphalerite occurs in galena areas it is in the form of rounded kernels, but in quartz and siderite areas it forms veins, patches, and connected areas of irregular outline. In many places the four principal minerals seem to be nearly contemporaneous, as in the alternating bands of them in the Blue Bird ores.

The age relations of the accessory minerals are not so clear, but as discussed in the general section on paragenesis (pp. 135–138) all of them with the possible exception of magnetite and pyrrhotite are believed to fall between the end members of the series discussed above.

SCOPE OF DESCRIPTIONS.

In the following sections emphasis is placed on the geologic conditions surrounding the principal lodes, the form and distribution of the ore bodies, and the ores. Only casual mention is made of development, production, and history; mine workings, with few exceptions, are not described. These features have been covered in previous reports, and the situation has not greatly changed. The reader is referred to the extensive record contained in the reports by Ransome and Calkins [33] and by Calkins and Jones [34] and in the papers by Hershey.[35] The development accomplished since the examination in 1904 upon which Professional Paper 62 is based is briefly outlined. A few new discoveries have been made, and these are more fully described. The deposits are grouped as in the earlier and more detailed report, and about the same order of subdivision is followed in order to facilitate reference from one report to the other.

[33] Ransome, F. L., and Calkins, F. C., The geology and ore deposits of the Coeur d'Alene district, Idaho: U. S. Geol. Survey Prof. Paper 62, 1908.

[34] Calkins, F. C., and Jones, E. L., jr., Economic geology of the region around Mullan, Idaho, and Saltese, Mont.: U. S. Geol. Survey Bull. 540, pp. 167–211, 1914.

[35] Hershey, O. H., Origin of lead, zinc, and silver in the Coeur d'Alene: Min. and Sci. Press, vol. 107, pp. 489–493, 529–533, 1913; Genesis of lead-silver ores in Wardner district, Idaho: Min. and Sci. Press, vol. 104, pp. 750–753, 786–790, 825–827, 1912; Origin and distribution of ore in the Coeur d'Alene (private publication), 1916.

LODES NEAR WARDNER.

SITUATION AND DEVELOPMENT.

The principal lodes occur within a zone well up on the rugged slope south of the valley of the South Fork of Coeur d'Alene River and nearly parallel to it. The zone extends from a point east of Milo Creek across the head of Deadwood Gulch to a point east of Government Gulch, a distance of about 2½ miles. It nowhere exceeds half a mile in width. The principal mines along this zone from east to west are the Bunker Hill and Sullivan, Last Chance, Sierra Nevada, Caledonia, Ontario, Senator Stewart, Silver King, and Crown Point.

The mines are opened principally by tunnels from Milo, Deadwood, and Government gulches and by the Kellogg tunnel, 1⅜ miles long, from the main valley west of Kellogg. From this tunnel there are direct connections with openings from each of the three gulches. The part of the Bunker Hill and Sullivan mine now principally worked is developed by a 1,000-foot shaft whose collar is on this level; the other mines are shallower.

PRODUCTION.

The production of the principal mines in the vicinity of Wardner, as shown in the following table, is particularly interesting when it is realized that the large output of the Senator Stewart, Caledonia, and Ontario came from deposits that were prospects at the time of Ransome's examination in 1904. At that time the area had been actively exploited for 19 years. Surely this is sufficient justification for a belief that other deposits of no mean importance will be discovered.

The magnitude of such a deposit as the Bunker Hill and Sullivan is shown not only by its total production. The Miller stope, on the March ore shoot above the thirteenth level, is yielding 8,950 tons to the vertical foot; the Cameron, on the same shoot above level No. 12, yields 7,000 tons to the vertical foot, and to the end of 1915 had produced 760,120 tons of ore. In 1904 practically all work was being done above the No. 9, or Kellogg tunnel level, and a yield of 6,000 tons per foot was obtained; below this level the average yield has been about 10,000 tons to the vertical foot. At present one-seventh of the Bunker Hill ore is obtained above the No. 9 level and six-sevenths below that level.

Production of principal mines near Wardner, Coeur d'Alene district, Idaho.

Bunker Hill & Sullivan Mining & Concentrating Co.

Year.	Crude ore.	Concentrating ore.	Ratio of concentration.	Silver.	Lead.
	Tons.	*Tons.*	*Per cent.*	*Ounces.*	*Pounds.*
Prior to 1904				8,338,632	621,326,000
1904	22,110	317,722	7.55	1,140,719	62,300,897
1905	36,570	342,470	6.40	1,556,252	79,608,970
1906	24,720	344,340	5.93	1,623,629	83,139,790
1907	9,290	313,040	4.46	1,415,367	71,440,320
1908	3,960	358,190	4.59	1,228,014	66,938,801
1909	2,150	341,420	5.05	1,179,412	65,466,571
1910	1,330	425,140	5.89	1,219,472	66,275,921
1911	1,790	464,750	6.74	1,250,058	68,740,718
1912	2,120	427,740	7.26	1,170,860	65,057,000
1913	179	436,643	7.48	1,209,354	70,935,265
1914		440,919	7.85	1,161,324	70,663,236
1915		454,636	7.59	1,298,284	74,584,741
	a 104,219	a 4,667,010		a 23,791,377	a 466,478,230

a The tonnage of ore is for the years 1904 to 1915, inclusive; the production for the life of the mine.

Last Chance mine.

1904 b	8,261	180,383	7.5	781,111	35,945,600
1905	8,137	162,058	7.8	625,688	31,315,600
1906	8,917	162,119	8.1	586,405	30,191,000
1907	5,928	158,872	8.3	473,473	25,195,800
1908	6,828	144,162	7.8	490,347	25,556,400
1909	7,952	176,198	6.7	756,020	35,553,600
1910	7,905	179,545	7.4	1,020,191	36,003,000
1911	5,322	194,928	9.2	1,047,291	32,507,000
1912	7,789	222,511	11.0	581,301	29,910,000
1913	6,268	263,632	13.7	501,245	27,227,794
1914 b	6,180	299,920	13.9	598,873	31,359,974
1915 b	2,812	133,368	13.0	243,655	14,110,024
1916 b	1,294	54,556	11.7	127,009	6,586,516
	83,593	2,332,252		7,832,609	361,462,308

b For the period prior to 1914 the figures are for the 12 months ending Aug. 31, for 1914 the 16 months ending Dec. 31, for 1915 the calendar year, and for 1916 the 5 months ending May 31.

Stewart Mining Co.

1910 c		55,637	8.22	325,037	8,429,906
1911	1,941	145,561	8.59	806,380	21,580,641
1912	5,838	179,505	8.08	1,202,272	29,167,762
1913	6,449	170,188	7.16	1,581,235	30,340,088
1914	6,137	189,615	6.11	2,648,422	45,656,486
1915	2,699	177,323	7.74	1,652,236	29,510,188
1916 d		33,151	9.02	199,607	3,421,546
	22,064	950,980		8,415,189	168,106,617

c Apr. 23, 1910, to Jan. 1, 1911.
d Jan. 1, to July 1, 1916.

Caledonia Mining Co.e

1914	4,042	18,420	5.67	572,196	6,866,040
1915	6,954	35,996	4.35	1,251,896	11,248,400
	10,996	54,416		1,824,092	18,114,440

e Production under present management. The mine began producing in 1909 and, according to the records of the county assessor, produced 18,880 tons of ore, valued at $988,453, prior to 1914. This is the only mine in the district in which copper is a constituent of noteworthy commercial importance. In 1914 the mine produced 246,000 pounds of copper, and in 1915, 668,000 pounds.

Production of principal mines near Wardner, Coeur d'Alene district, Idaho—Con.

Ontario Mining Co.[f]

Year.	Crude ore.	Concentrating ore.	Ratio of concentration.	Silver.	Lead.
	Tons.	*Tons.*	*Per cent.*	*Ounces.*	*Pounds.*
1911	58	2,130	8.94	11,437	266,686
1912	1,588	43,593	7.13	277,744	9,674,256
1913	1,054	71,838	9.69	353,742	8,654,352
1914	2,350	86,460	10.48	491,874	11,461,952
1915	1,668	79,551	11.05	444,884	9,135,600
1916	252	28,338	9.30	164,328	3,471,198
	6,970	311,910		1,744,009	42,664,044

f Data cover the life of the mine.

Sierra Nevada Consolidated Mining Co.[g]

1913		37,024	9.54	74,594	3,286,530
1914		77,457	6.37	356,631	12,083,856
1915		48,832	12.04	142,433	4,094,786
		163,313		573,658	19,465,172

g Production under present management. The mine was idle for a number of years prior to 1913.

GEOLOGIC RELATIONS.

In a discussion of the geologic relations of the Wardner ore bodies too much credit can not be given to the vast amount of work done by O. H. Hershey, geologist for the Bunker Hill & Sullivan Co., and the several geologists who at different times have studied the deposits incidentally to litigation. Few mining companies have more complete geologic records than the principal company operating in this district. A glass model with structure shown on every level and vertical sections at 100-foot intervals was prepared and is kept up to date by Mr. Hershey. In this way the detailed structure of the large block of ground honeycombed by workings is intimately known, and any inconsistencies in the correlation of faults or lodes are soon discovered and the errors corrected. By working agreement or by overlapping management Mr. Hershey keeps similar records of most of the adjoining mines. These data were available to the writers during their examination of the district and contributed more largely than their own work to their present knowledge of the deposits. To give credit on particular points is impossible, but it may be said frankly that the recent examination contributed very little that is new on the details of structure. By no means all the structural details, particularly those of the surface and levels above No. 9, were checked by the writers, but in regard to the many that were there is little room for disagreement. Only on the interpretation of the facts are noteworthy differences of view held. These differences are stated freely, not in a spirit of criticism but solely in the hope of furthering our knowledge of ore genesis. Instead of nine distinct periods of faulting caused by regional forces operating differently and in different directions and ten separate stages of mineralization, as held by Hershey,[36] the writers recognize not over two distinct periods of faulting—those common to the entire region (see pp. 11–14 and fig. 3)—and only one general period of mineralization. The slicing of the Wardner area into a great number of thin slabs of steeply dipping beds, with minor cross faulting, is conceived to have been incident to readjustments along the curved plane of the Osburn fault during a long period in which the 12-mile relative displacement of its sides was accomplished. Many of the structural complexities are elucidated by recognizing the area as part of the east limb of a broad anticline crossed in the vicinity of the mines by a

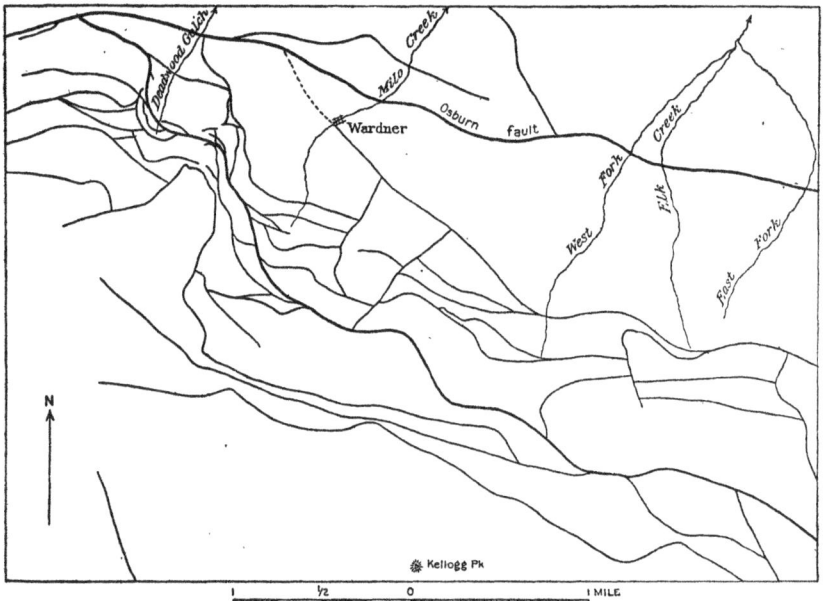

FIGURE 3.—Traces of faults in the Wardner area.   (After Hershey.)

sharp east-west syncline and these folds traversed obliquely by parallel faults, most of which have notable horizontal components of displacement. Thus the faults of the Wardner area, with the possible exception of certain fragmentary ones, such as the Good Luck, Forest Bell, and Monument, which have a prevailing north-south strike, are believed to date from the third general period of structural disturbance.

The ore of the Bunker Hill and Sullivan, Last Chance, and Sierra Nevada mines lies principally in the Revett formation, but the Caledonia, Senator Stewart, and Crown Point are in the Burke. With increased depth the Bunker Hill and Sullivan ore bodies are gradu-

[36] Hershey, O. H., Genesis of lead-silver ores in Wardner district, Idaho: Min. and Sci. Press, vol. 104, pp. 750–753, 786–790, 825–827, 1912.

ally passing into steeply dipping Burke beds. The ore bodies of the Bunker Hill and Last Chance mines are confined to a block of ground bounded by the Sullivan, Buckeye, and Iron Hill faults, shown on Plate VIII. Within this slablike block, approximately 6,000 feet long by 1,000 feet or less wide, which strikes N. 45° W. and is inclined about 40° SW., the ore bodies are irregularly distributed, although in a general way they occur near the footwall, or Sullivan fault. In no place has the ore been found to extend beyond this footwall, but in some of the fissures of the Last Chance mine it extended beyond the Buckeye fault. Many minor faults, some of which displace the lodes, traverse this mass of quartzite, but none have sufficient throw to break the general continuity of the block. Important among them are the Upper Cate, Lower Cate, Flint, Dull, Last Chance, Ollie McMillin, and Jackson faults. A general idea of the faulting within the block is shown in Plates VIII and IX, comprising a geologic map of No. 9 level and two vertical sections through the Bunker Hill and Sullivan mine, kindly furnished by Mr. Hershey. The Lower Cate fault passes through the March ore shoot on many levels but has caused no serious displacement of it, although the combined movement on the Upper Cate and Lower Cate faults offsets the Jersey fissures about 800 feet to the west on the south side. On level No. 12 the two parallel Jersey fissures are about 100 feet apart and the east one is offset nearly 80 feet by the Last Chance fault, which is at right angles to their course, but the west one is not displaced. The Blue Bird seams are ribbon-like in structure, indicating recurrent opening of the fractures during their development. It is also noteworthy that the deposits exemplifying the three principal types of mineralization recognized by Hershey— the Bunker Hill, Blue Bird, and Jersey—have different strike and a slightly different assemblage of minerals. All these facts show clearly that faulting movements of no small importance progressed concurrently with mineralization.

The Caledonia vein occupies a northeast fissure centrally situated in a mass of Burke quartzite bounded on the north by the Osburn fault and on the south and west by the New Era fault. The vein is extremely crooked and is cut by numerous faults of minor displacement; it dips southeast. On the 950 level the vein terminates against a fault zone which has not been recognized on the surface or elsewhere in the mine. This vein is of considerable interest because it lies well to the north of the so-called Wardner footwall, or Sullivan fault, which was originally believed to be the northern limit of commercial mineralization.

The Senator Stewart vein has about the same strike and dip as the Caledonia and lies west of it in a Burke area bounded by the Osburn, Lower Cate, and New Era faults. The vein is set down on the west by three strike faults, so that more than one segment appears on several levels. At a depth of 735 feet below the outcrop it is cut off by the New Era fault, which dips about 35° W. The faulting movement dragged fragments of the ore in commercial quantities locally to a distance of 80 feet up on the fault plane. On the northeast the vein is cut off by the Osburn fault, and on the southwest by the Lower Cate fault, also called the Ontario footwall. The position of the New Era fault between this vein and the Caledonia suggests that they may be parts of the same deposit. Further, Hershey [37] has suggested that the Caledonia is probably a portion of the Sierra Nevada, which lies well to the south and is cut off on the northeast by the Cate faults. It is perhaps a reasonable working hypothesis that the Senator Stewart is a central segment of a vein which once included the Caledonia and the Sierra Nevada. It is also possible that the Crown Point and Silver King veins, well to the west, may be faulted parts of the Senator Stewart, although this seems less likely. In depth the west end of the Senator Stewart vein passes into Ontario ground and is worked by the Ontario Co. The problem of locating the faulted ends of the Stewart vein, an exceptionally rich deposit cut off sharply on three sides by faults, is well worth attempting. The limited time available to the writers for the examination of local problems makes it impossible for them to do more than outline this problem, but in approaching its solution the investigator should be guided by the facts that much, perhaps most, of the displacement on the Osburn fault probably preceded the vein formation, and that in the Bunker Hill and Sullivan mine the net result of movements on the Upper Cate and Lower Cate faults is an offset of about 800 feet to the west on the south side of the Jersey fissures, which have similar strike, dip in the same direction, and are of the same general type as the Stewart vein. Postmineral movement on the Lower Cate fault has been more pronounced than on the Upper Cate, so that the chances of finding another segment of the Stewart vein would seem to be greatest in the block between these two faults at some point west of the vein as now known.

The Crown Point and Silver King veins dip northward and are cut off by the Osburn fault, which dips about 50° S. In the former much work has been done on the assumption that the fault might possibly be older than the ore, but this work has conclusively proved that an important part of the movement on the fault is younger. West of the Crown Point in the same general relation to the Osburn fault is the Black Hawk vein.

It is obvious that much of the faulting in the Wardner area is older than any of the lodes and that some is clearly younger, while evidence

[37] Hershey, O. H., Genesis of lead-silver ores in Wardner district, Idaho: Min. and Sci. Press, vol. 104, p. 789, 1912.

of faulting concurrent with ore deposition is equally conclusive.[38] The development of a 12-mile displacement of the earth's crust must be a slow phenomenon, probably characterized by innumerable advances separated by periods of accumulating stress. It is reasonable, therefore, to conceive that the general period of movement along the Osburn fault exceeded in duration that of ore deposition, even though concurrent with it in the main.

South of the Osburn fault ore has also been found in the Alhambra mine, on Elk Creek, in the Newland and Revett formations, here separated by a mineralized fault. The ore solutions clearly followed the fault zone and replaced the rocks on both sides.

North of the Osburn fault there are several minor deposits which have been rather fully described by Hershey.[39] They are zinc, zinc-lead, and gold-copper veins, mostly in the Prichard formation; the East Caledonia, a lead deposit, is in the Burke. The Prichard here is divisible into two slate members separated by a quartzite member about 75 feet thick. Most of the prospects are in the middle and upper members, but a few occur in the lower member. The Evolution vein, carrying galena, pyrrhotite, and some sphalerite and copper in a ferruginous calcite gangue, is probably lower in the Prichard than any of the others. Some faulting occurs in the area north of the Osburn fault, but it is vastly less prevalent than in the area south of the fault, as shown by the almost unbroken outcrop of middle Prichard quartzite.

Dike rocks are fairly abundant in the Wardner area. North of the Osburn fault both light-colored and porphyritic dikes and fine-grained basic dikes are abundant. In most places they are so greatly decomposed that microscopic examination adds little to field determinations. The acidic dikes are probably near monzonite porphyry in composition, and the basic dikes are principally lamprophyres. Hershey has pointed out that the acidic dikes are older than the ore deposits and the basic dikes are younger. In places along the Osburn fault crushed lamprophyre appears in the gouge or in the rock adjacent to it. The dikes are in no place known to be accompanied by noteworthy contact metamorphism.

FORM AND DISTRIBUTION.

The general distribution and structural relations of the lodes have been discussed in the previous section. Here the form and distribution of the ore bodies within the lodes will be considered. Plates X and XI give a general idea of mining operations and the distribution of ore bodies in this area.

[38] Hershey, O. H., Genesis of lead-silver ores in Wardner district, Idaho: Min. and Sci. Press, vol. 104, p. 750, 1912.
[39] Hershey, O. H., Origin and distribution of ore in the Coeur d'Alene (private publication), 1916.

In considering the deposits it is convenient to recognize three sets of lodes, all of which are believed to represent one general period of mineralization. The types of lodes here recognized have been designated by Hershey[40] the Blue Bird, Jersey, and Bunker Hill. Hershey distinguishes several other phases of the mineralization, but as they are either closely related to these in character or grade into them they will not here be considered separately.

Blue Bird lodes.—The Blue Bird lodes are made up of narrow veins of ore minerals from a fraction of an inch to 2 feet or more in width. In most places the intervening quartzite, which contains disseminated siderite and pyrite but only locally any sphalerite or galena, is mined, the value of the material depending on the closeness of the spacing of the lead-bearing seams. The lodes strike about N. 70° W. and dip 30°–80° SW. This phase of the ore deposition was not well illustrated in stopes accessible at the time of the examination by the writers, but Hershey's statement concerning it is full and detailed and is quoted in part:

> The Blue Bird solutions primarily circulated along cracks in the quartzite, possibly attacking the walls somewhat; but it is true throughout the mine, with one important exception, that fissure filling and not metasomatic replacement of the quartzite was the predominant characteristic of the Blue Bird stage. * * * Some of them [the cracks] are only bedding-plane slips, though most of them cut the bedding at small angles. A characteristic feature, especially in the Last Chance mine, is the presence of many veinlets of ribbon quartz. The quartz is of a fine-grained, hard, slightly bluish type. A rather fine grained pyrite, a rather light colored zinc blende, some galena, and a little siderite tend to be distributed along fine parallel lines, giving the rock a banded appearance.

Four principal ore shoots, or sections in which the lead-bearing veinlets are so closely spaced as to warrant mining, may be distinguished. Descriptions by the same author follow:

> The most eastward, the Hatton, has been stoped from the surface to the Cedar level of the Bunker Hill mine and has produced about 160,000 tons of ore. A great group of more or less connected stopes represents the Roberts shoot, which extends from the surface to a vertical depth of about 700 feet and has probably produced about 600,000 tons of ore. The heart of the Fan ore shoot was one of the most persistent and straightest ore bodies in the district and is represented by a nearly continuous series of large filled stopes in the Last Chance mine and a new stope in the Bunker Hill mine. It extends from the base of the oxidized zone to a vertical depth of 1,000 feet and has produced probably over 100,000 tons of ore. The most westward shoot, the Edna, is exceedingly irregular but can be traced from the base of the zone of oxidation to a vertical depth of 1,300 feet and to that depth will produce probably 150,000 tons of ore.

In the east end of the Bunker Hill and Sullivan mine the Blue Bird fractures are steeper than a large siderite-pyrite vein called the

---

[40] Hershey, O. H., Genesis of lead-silver ores in Wardner district, Idaho: Min. and Sci. Press, vol. 104, pp. 787, 788, 825, 1912.

Motor seam. The solutions rising along them were diverted along the Motor seam and transformed it into ore from the place of intersection to the surface. On the No. 9 level the filling is a mass of nearly barren siderite and pyrite, but above a level 50 feet higher more than 700,000 tons of excellent ore had been mined prior to 1912.

The Buckeye type of Hershey " is a product of the Blue Bird stage that is confined to the vicinity of the Buckeye fault " and is best developed in an area in the east end of the mine, where fractures are widely spaced and disseminated sphalerite, pyrite, and galena are associated with a little chlorite and sericite. Specimens examined for Mr. Hershey by A. C. Lawson contained also an occasional crystal of tourmaline, but this mineral has not been recognized in sections examined by the writers.

To the west the Blue Bird type gradually loses its pyritic content and consists of galena, quartz, and a little sphalerite as a cement in brecciated quartzite (" Ontario type ").

Bunker Hill lode.—The Bunker Hill lode contains the most productive ore bodies in the Wardner area, having produced up to 1912 over 2,584,000 tons of ore and at present affording more than three-fourths of the tonnage of the district. The ore differs from the average Blue Bird ore in containing much less pyrite, but it is similar in mineral assemblage to the Blue Bird ore found in the west end of the mine. Structurally the principal ore body, the March shoot, is roughly related to a triangular block of quartzite bounded by the Sullivan, Flint, and Upper Cate faults. It has been followed from the surface to the No. 14 or lowest level opened, a distance on the pitch of the shoot of more than 4,000 feet. Below No. 9 level the relation to the three faults is not very close. The Sullivan fault dips more steeply than the ore, on No. 10 level being 300 feet in the footwall, and on No. 11 level a drift extending nearly 600 feet into the footwall has not cut it. On lower levels it has not been exposed. The Flint fault, which strikes northeast and dips northwest, holds closer to the ore shoot. On levels Nos. 11 and 12 most of the ore is east of the line of the Flint fault, but on No. 13, as on No. 9, and again on No. 14, the fault forms the footwall. A bend in the Upper Cate fault arches over the March shoot from the surface to the lowest level opened and seems to have been the prime factor in directing the solutions. This inverted trough pitches southwest with the shoot, and nowhere does the ore depart far from it, although on No. 14 level, only partly opened at the time of visit, there is some suggestion that the Upper Cate fault is becoming flatter than the March shoot. The Lower Cate cuts through the ore shoot on levels 10, 11, and 12 but does not displace the ore because the direction along which it moved is here parallel to the pitch of the shoot. The relations below No. 9 level are shown in Plate XII, prepared by the writers, but with certain areas not safely accessible to them filled in from a glass model prepared by Mr. Hershey and others.

In several places branches from the continuous March shoot extend to the side and upward. In reference to the principal ones Hershey writes as follows: [41]

> From a point below Bunker Hill No. 8 level to No. 12 level an arm extending along the Ikey seam southeastward from the March shoot had produced about 248,000 tons up to June 1, 1910. Another Bunker Hill ore shoot came to the surface in the Scrafford open cut; the original discovery of lead in the district was made on it. Before the Lower Cate faulting it was merely a northwest wing to the March shoot, with which it merged in depth. The total production has been about 378,000 tons. The East Tyler is another Bunker Hill shoot.

Jersey veins.—The Jersey " fissures," as they are known in the district, may more accurately be termed veins than lodes, although replacement has played a considerable part in their formation. In general aspect they approach typical fissure fillings more closely than any other of the lead-silver deposits. They also differ from the Bunker Hill and Blue Bird lodes, although containing the same minerals, in having quartz instead of siderite as the predominant gangue constituent.

These veins include the Jersey and its faulted segment the Barr, the N and its continuation the Francis, and the Sierra Nevada, Senator Stewart, and Caledonia, probably faulted parts of the same vein. The veins in general strike east-northeast and dip south-southeast. They are much more disturbed by faulting than the veins of either of the other groups, but it is not impossible that this is merely due to their course, which is transverse to the direction of recent faulting, rather than to a marked difference in the age of their formation.

The Jersey vein, nearly worked out at the time of the writers' examination, is described by Hershey [42] as follows:

> The Jersey fissure, which has been extensively mined in the Last Chance mine and is now being stoped in the Bunker Hill mine, originally extended from the vicinity of the Bunker Hill fault to beyond the Buckeye fault. Starting at the surface in upper Burke strata, with depth it passes the Tyler fault into the hard pyritic lower Revett. * * * The main ore shoot averages several hundred feet in length and is clearly traceable from Bunker Hill No. 9 level to the Last Chance O level, 2,000 feet on the dip. It has a straight course, almost due south, which is practically straight down on the dip of the vein as a whole. In ascending, the main ore shoot gradually improved to a point a little below where it passed from lower Revett into upper Burke quartzite, and thence there was a gradual decrease. This was connected with the development of a great concave curve, which becomes of less radius as height is attained. * * * The production of the Jersey fissure when it has been exhausted down to Bunker Hill No. 9 level will considerably exceed 400,000 tons.

---

[41] Genesis of lead-silver ores in Wardner district, Idaho: Min. and Sci. Press, vol. 104, p. 825, 1912.

[42] Op. cit., pp. 788–789.

Parallel to the Jersey vein in a general way and about 200 feet northwest of it on No. 9 level is the N, a very crooked vein with two principal ore shoots, now worked out, which produced up to 1912 about 100,000 tons of ore.

On the northeast both the Jersey and the N are cut off by the Upper Cate fault. Between this and the Lower Cate fault the veins have not been recognized, but about 800 feet to the southeast, below the Lower Cate fault, from above level No. 11 down to the 14th or deepest level, two similar veins with similar strike and dip and similarly spaced are being mined. The relations are shown in Plate XII. Here the southeast vein is known as the Barr and the other as the Francis. In their western parts the Barr and Francis veins are parallel between levels Nos. 11 and 12, but between Nos. 10 and 11 the Barr is steeper than the Francis. On levels 11, 12, and 13 the two veins merge northeastward, but both become barren of lead before reaching the point of junction. From level No. 12 in the vicinity of the Dull fault a shoot 60 feet long and one set wide begins at a point 150 feet beyond the end of the main Francis shoot and continues thence to the March shoot.

The Sierra Nevada vein as described by Hershey [43]

has the form of an anticline whose axis strikes S. 45° W. and descends 85 feet in 800 feet, until it reaches the Oakland fault, after which it pitches more steeply. * * * The main ore shoot occupied the nearly flat portion near the axis of the anticline, with some lenses probably 6 feet thick. * * * The vein is extremely crooked and is cut off on the northeast by the Cate faults.

The Caledonia vein, situated in an area of Burke quartzite north of the Cate fault zone and about 2,500 feet from the Sierra Nevada, is also crooked and is cut by numerous faults of small displacement. Ore occurs in one principal shoot about 400 feet long and several small and poorly defined shoots along the same zone northeast and southwest of it. At the time of visit stoping was being done at various places between the 500-foot and 950-foot levels.

The Senator Stewart vein, about 2,000 feet west of the Caledonia, beyond the New Era fault, has similar strike and dip and is made up of the same minerals. Ore within it was widely distributed, occurring partly in the Ontario mine, against the Lower Cate fault on the southwest, thence to the Osburn fault on the northeast and down to its termination against the New Era fault. The ore follows a narrow fissure whose formation was accompanied by extensive crushing; locally the vein is as much as 30 feet wide.

The Silver King and Crown Point veins, although dipping northward, should probably be included in the Jersey group. In the

[43] Op. cit., p. 789.

Silver King bunches of ores were found in the shattered quartzite that forms the hanging wall of the Osburn fault. The Crown Point vein above the point where it is cut off by the Osburn fault contained considerable ore in an irregular sheet from 1 to 20 feet thick which dipped 20°–30° NE.

THE ORES.

In considering the ore deposits of the Wardner area, three principal types may be recognized to advantage, even though all contain the same minerals deposited in the same sequence and differ only in the relative amounts of the minerals. In the Jersey veins quartz is the dominant gangue, and in the Bunker Hill and Blue Bird lodes siderite is vastly preponderant. There is much pyrite in the Blue Bird and little in the Bunker Hill. Tetrahedrite, presumably argentiferous, is most abundant in the Jersey veins, but in different members of the group it has a wide range in relative amount. In the Blue Bird lodes of the east end of the area more sphalerite occurs than elsewhere, and this is the area in which disseminated sphalerite and a little galena are most abundantly developed. Also in the east end of the area some lodes are made up almost exclusively of siderite and pyrite.

The Bunker Hill ores consist essentially of galena in a siderite gangue, with quartz, sphalerite, pyrite, tetrahedrite, and chalcopyrite in accessory amounts. The ore is typically a replacement of quartzite, and the boundaries of the ore bodies are exceedingly irregular in details. Ransome's description of the ore [44] applies equally well to the deeper levels now opened. He says:

The best ore from such stopes [the March stope on No. 9] consists of rather fine grained masses of galena with subordinate siderite. This grades into ore in which the siderite exceeds the galena, and this into barren quartzite. The ore is principally a replacement of the Revett quartzite, but the replacement is closely connected with fissuring, and some of the galena was deposited in open spaces. A typical mass of ore of moderate richness shows a base of pale-brown siderite traversed by countless reticulating veinlets of galena. In the poorer ore these veinlets are fairly distinct and show only slight metasomatic enlargement of the original cracks. But in the richer ore they coalesce into bunches of solid galena. It is plain that the quartzite was first fissured and replaced by siderite. Subsequently the siderite was shattered and solutions deposited galena in the intersecting fissures and, by metasomatic replacement, in the siderite of their walls. Where the little fissures were particularly numerous and the other conditions for deposition favorable, the siderite has been wholly replaced by galena. But the process was not entirely so simple. Some of the galena is traversed by stringers of siderite, and by veinlets of galena of different (usually finer-grained) crystallization from the mass of the ore. Moreover, here and there galena directly replaces the quartzite. It is evident, therefore, that there has been some recurrence of conditions favorable

[44] U. S. Geol. Survey Prof. Paper 62, p. 162. 1908.

to the deposition of galena and siderite and probably also of quartz, pyrite, sphalerite, and other metasomatic minerals. In the Bunker Hill stopes quartz and pyrite are usually most conspicuous in the transition zone from ore to country rock.

According to the observations of the writers, pyrite in the main was deposited with the siderite and occurs in isolated crystals or groups of crystals scattered through its mass. When the siderite was fractured most of the cracks passed around the pyrite crystals, apparently because they were stronger than the crystals of carbonate, but in many places the pyrite is fractured and crossed by seams of the galena. Sphalerite also was an early mineral to form, and no second generation of it was observed. It occurs characteristically as rounded kernels in galena areas, but in many places appears in irregular replacement contact with the siderite. It seems that when the lead was deposited it replaced the carbonate more readily than the zinc sulphide, thus working around it and attacking its irregular protuberances. Quartz followed the siderite and pyrite in the main and clearly replaces these minerals. In turn it is replaced by galena, which traverses it along fractures with enlargements irregularly distributed, and probably also by sphalerite. It is clear that the sequence of mineral deposition in the Bunker Hill ores, as seen in a large way in the stopes and confirmed by polished surfaces, is siderite, with accompanying or slightly later pyrite, quartz, sphalerite, and galena, with a recurrence in very subordinate amounts of each of the minerals except possibly sphalerite.

The Blue Bird ore differs from the Bunker Hill ore in consisting largely of fissure fillings having more quartz, much more pyrite, and locally more zinc. The ore of this group of lodes, however, differs markedly in different places. Locally, as in the Last Chance mine, quartz exceeds siderite, but elsewhere siderite is preponderant. Pyrite is notably scarce in the west end of the mine, but abundant elsewhere. Where pyrite is sparse in the lodes it is nearly absent as a dissemination in the quartzite, and where sphalerite is abundant in the lodes it also occurs in disseminated form. Accompanying the lodes are innumerable small veinlets, bunches, and blebs of pyrite, galena, and sphalerite. These narrow veinlets are mostly isolated from important lodes in the east end of the area, but where they intersect earlier siderite bodies they open into large commercial shoots, as in the Motor seam. In many places the veinlets that make up the lode system are too narrow and too widely spaced to permit mining. Characteristically they have a ribbon structure due to bands of quartz alternating with bands of siderite, pyrite, galena, and sphalerite. Locally all the sulphides occur in the same band, but in most places bands are made up predominantly of one or two of them. Bands of pyrite commonly separate bands of quartz. The galena occurs mostly in siderite layers, but in places it cuts across both quartz and siderite. Exposures in the Flood level show the order of deposition to be siderite, quartz, pyrite, sphalerite, and galena, which seems to be the characteristic sequence. In places such ore is traversed by veinlets of quartz and pyrite. Where both siderite and pyrite occur in the same veinlet pyrite is either of the same age as the siderite or younger.

The Blue Bird lodes afford abundant evidence in their characteristic banding of repeated reopening of the cracks along which the solutions traveled.

The Jersey ores characteristically contain considerable quartz, which replaces siderite formed along fissured and sheared zones. The sulphides, chiefly galena, replaced both the siderite and the quartz from cracks which are irregularly distributed and in many places very closely spaced. Sphalerite, tetrahedrite, and chalcopyrite occur in patches, rounded kernels, and veinlets which cut the siderite and quartz and are themselves cut, at least in many places, by galena: Sericite occurs locally as nests in the quartz. It is probably the greater amount of tetrahedrite in the Jersey ores that gives them a silver content unusually high for the district. In the Barr vein the siderite stage was less pronounced than in the Francis, so that the ores of this vein consist more largely of quartz and galena. Eastward the Francis and Barr converge, but from the relations at their point of union it is impossible to recognize any evidence showing a difference in age of the two veins. The quartz from both merges and continues unbroken beyond; certainly they are essentially contemporaneous. On levels 12 and 13, however, the Last Chance fault (Pl. XII) offsets the Barr vein as much as 100 feet but does not affect the Francis, only a little more than 100 feet farther north. This can not be explained by strike and dip relations, because the two veins are essentially parallel at this place and the fault is transverse to them. It seems most reasonable to conclude that the Barr fissure is older than the principal displacement on the fault and the Francis fissure younger but that the mineralization of both is younger. Similar age relations may exist between the Jersey and N, southwest of the Cate fault zone, and the Francis and Barr, their continuations respectively to the northeast of the Cate faults and 800 feet southeast.

The Caledonia vein is similar to the Barr and Francis but contains more tetrahedrite with consequent increase in copper and silver. Indeed, this is the only lode in the district that yields copper in noteworthy quantity, although in most of the other Jersey veins copper is reported by the smelters. The ore is primarily valuable for its silver and lead. Galena, clearly formed later than the tetrahedrite, which it traverses in seams and rounded kernels of which it incloses, is confined to central portions or shoots in the tetrahedrite-bearing ore. Chalcopyrite occurs with the tetrahedrite. The dominant sequence of formation of the primary minerals is siderite, probably accompanied by pyrite, quartz, sphalerite, chalcopyrite and tetrahedrite, and galena. In the Caledonia vein oxidation has progressed further than elsewhere in the Wardner district. Much iron and manganese oxides, cerusite, native silver and copper, a little pyromorphite, covellite, azurite, malachite, and massicot occur at many places between the surface and the 950-foot or lowest level to which ore has been followed. Argentite is rare, and neither anglesite nor cerargyrite was observed. In oxidation the siderite is leached, leaving a skeleton of quartz that affords many cavities filled or lined by the secondary minerals. Cerusite in beautiful crystals as much as 1½ inches in length is very abundant. In general pyromorphite and massicot occur near the surface, native silver on the intermediate and lower levels, and native copper in the lower 200 or 300 feet. It is noteworthy, however, that pyromorphite in well-formed crystals attached to fresh galena has been found on the 900-foot level and massicot down to the 850-foot level.[45] Pyromorphite was found developed in well-formed crystals on the mine timbers of an abandoned drift on the 200-foot level by Mr. Rush White, mining engineer, of Wallace, Idaho. The explanation for the deeper oxidation in the Caledonia mine than elsewhere in the district seems to be the large number of minor faults transverse to the vein.

The Senator Stewart ore is similar to the Caledonia, but contains less copper and silver and shows only slight oxidation. In view of the probable fault relations between the two veins, it is noteworthy that silver is relatively much more abundant in the lower Stewart levels than in the upper ones. In the upper levels there was less than 1 ounce of silver to 1 per cent of lead, and in the lower levels considerably more than 1 ounce. In the vein a little barite occurs locally.

The Sierra Nevada, Crown Point, and Silver King ores all show a high silver ratio, and quartz is an abundant constituent of the gangue. In the Sierra Nevada ore oxidation similar to that in the Caledonia is extensive.

LODES NEAR MULLAN.

SITUATION AND DEVELOPMENT.

Two principal deposits near Mullan are of major importance—the Morning, 2 miles north-northwest of Mullan, and the Gold

---

[45] Specimen in collection of E. V. Shannon, Y. M. C. A. Building, Kellogg, Idaho.

Hunter, 1 mile northeast of Mullan. Other veins that have afforded some production include the Alice, on Ruddy Gulch, 2½ miles west of Mullan; the You Like, 1,000 feet south of the Morning and parallel to it; and the Carbonate Hill, on Willow Creek about 2 miles above its mouth.

The Morning lode is worked through a tunnel 11,000 feet long, which also cuts the You Like. The tunnel corresponds to the 800-foot level, 2,200 feet vertically below the outcrop of the vein, and from it a shaft extends 1,050 feet to the 1,850-foot or deepest level. The Hunter lode is worked through a vertical shaft 800 feet below the No. 6 tunnel, an adit 4,230 feet long which cuts the lode 1,600 feet below its apex. The level numbers are carried from tunnel No. 5, which is 400 feet higher, so that the deepest level is known as No. 12. The Alice vein is developed by a main adit, the 100-foot level, and a shaft extending 600 feet deeper. Development on the Carbonate Hill consists of about 3,000 feet of tunnel and drifts.

PRODUCTION.

The production of the principal lodes is shown in the accompanying tables. Records for the Alice are not available, but its production has probably not been much in excess of $100,000. Details of production are shown only for the years since the date of Ransome's report. The extent to which zinc has entered into the output of the Morning mine since 1911 is particularly noteworthy. This is due to a modification in concentrating methods, to an increase in zinc with depth, and to an increased extraction from the west end of the lode, in which zinc is much more abundant than in the east end.

Production of the Morning mine, 1895–1916.[a]

Year.[b]	Crude ore.	Concentrating ore.	Ratio of concentration.	Silver.	Lead.	Zinc.
	Tons.	*Tons.*	*Per cent.*	*Ounces.*	*Pounds.*	*Pounds.*
1895–1905				4,467,653	234,360,598	
1906	14	248,617	8.6	415,976	25,812,200	
1907	48	332,452	8.7	554,450	34,940,600	
1908	279	147,321	8.8	240,409	14,826,000	
1909	118	359,257	9.1	531,298	33,667,400	
1910	685	342,615	10.1	432,760	26,746,600	
1911	13,373	347,827	10.4	624,765	35,180,040	497,480
1912	27,270	374,030	10.4	726,686	44,710,880	2,375,800
1913	23,199	374,401	9.3	671,641	42,489,860	5,960,540
1914	26,042	299,458	7.7	714,064	42,985,740	6,030,420
1915	26,313	223,947	7.3	685,949	37,967,360	4,837,420
1916	21,688	135,534	7.5	526,310	23,471,088	3,638,172
	139,029	3,189,029		10,591,961	497,148,366	23,339,832

a 1887–1895, no record. The production of the You Like lode is not separated from the Morning lode in records available to the writers, but its contribution has been comparatively very small.

b Figures for 1906 are for 10 months ending Aug. 31; 1907–1913, 12 months ending Aug. 31; 1914, 16 months ending Dec. 31; 1915, the calendar year; 1916, 5 months ending May 31.

Production of the Gold Hunter mine, 1907–1915.[a]

Year.	Ore extracted (tons).	Gross value.	Year.	Ore extracted (tons).	Gross value.
1907......................	49,453	$172,616	1912......................	114,803	$263,420
1908......................	37,667	206,663	1913......................	114,000	489,148
1909......................	61,979	149,014	1914......................	110,543	421,318
1910......................	82,481	308,360	1915......................	118,764	614,590
1911......................	86,191	366,957			
				775,881	2,992,086

a Production prior to 1907, as given by Ransome, 160,249 tons, containing 606,475 ounces of silver and 9,563,642 pounds of lead. Later production from statements filed with the county assessor and compiled by the Wallace Miner, issues of Dec. 30, 1915, and May 11, 1916.

GEOLOGIC RELATIONS.

The lodes near Mullan occur higher in the stratigraphic series than any other of the principal deposits in Shoshone County. The Morning lode, in which zinc equals or exceeds lead in many stopes, occurs in the Revett formation; the other large zinc deposits are in the Burke or Prichard. The Gold Hunter lode is in the Newland ("Wallace") formation, a few thousand feet higher stratigraphically than the other principal lead-silver lodes, which traverse the Burke and Revett quartzites. The Carbonate Hill lode is in St. Regis quartzite.

The Morning and You Like lodes follow well-defined shear zones which strike west-northwest and, where not vertical, dip steeply northeast, forming a large angle with the bedding of the quartzite, which in general dips east. Calkins and Jones[46] observed that

The bedding of the country rock near the veins is steeply inclined and remarkably variable in strike, as if the strata had been raised to a nearly vertical position by pressure in one direction and afterward crumpled by pressure acting nearly at right angles to that direction.

The quartzite near the lode is distinctly sericitic, and locally a little chlorite is developed along the innumerable parallel planes of the shear zone. No igneous rock was seen in the part of the mine now being worked, although a small lamprophyre dike is known to occur in the eastern part on the upper levels. To the west, in the adjoining Ivanhoe and Star ground and on the course of the same shear zone, a monzonite porphyry dike at least 25 feet wide is exposed in the face of a 1,600-foot west drift from a prospecting tunnel about 2,500 feet long that starts in Grouse Gulch.

Evidence of displacement due to the shearing movement that formed the ore zone is not clear, owing to the similarity of the quartzite beds of the Revett formation. A distinct gouge from 3 inches to 2 feet thick follows the ore, in most places near the south wall, and strongly suggests postmineral movement, although so far as could be observed the walls on opposite sides of the ore body match closely.

46 Calkins, F. C., and Jones, E. L., jr., Economic geology of the region around Mullan, Idaho, and Saltese, Mont.: U. S. Geol. Survey Bull. 540, p. 180, 1914.

The gouge is characteristically tough and sticky, although locally it has a granular texture due to an abundance of minute fragments of ore and quartzite. The gouge seam continues as a fairly distinct feature beyond the limits of the ore bodies.

Fractures transverse to the lode are nowhere important except in the eastern part of the mine, where a north-south fault brings the Newland and Revett formations into contact on the surface. In the lower workings this fault must be well beyond the limits of known ore bodies. On the 1,650-foot level a drift extending 700 feet beyond the easternmost slope has not cut the fault, although a number of minor slips possibly related to it cross the shear zone.

The Hunter lode traverses a block of the greenish slate of the lower Newland brought against the Revett quartzite on the south by the White Ledge fault. "Another fault bringing middle Newland down on the north is shown by surface indications to cross the ridge about 400 feet north of the lode." [47] This fault crosses the main tunnel 140 feet north of the shaft and continues downward through the mine. Between these faults but south of the lode is another fault parallel to them which has a small downthrow on the north. The lode is much more complex than the Morning lode, owing to the greater fissility and more widely distributed fracturing of the inclosing rock. In general the lode strikes about west-northwest and dips about 80° S. The lode adjacent to it dips steeply south. The fault that crosses the tunnel north of the shaft seems to terminate the ore body on the east and has been followed down through the mine to the lowest level. It is not well exposed in present workings but where seen strikes about northwest and dips at a moderately steep angle southwest. Thus the line of intersection of the fault and the lode slopes westward. On the main tunnel or 400-foot level the fault is cut 140 feet north of the shaft, on the 600-foot level about 25 feet south of the shaft, and on the 800-foot level about 120 feet south of the shaft. On the 1,200-foot level there is a wide zone of intense shearing, and it is impossible to identify the particular gouge which represents this fault. The only place where the ore and fault were seen in contact was on the 600-foot level at a point 175 feet east of the shaft. Here there is a strong suggestion that unbroken ore seams turn into the plane of the fault in a zone 3 feet wide, about two-thirds of which is made up of ore minerals. Near the footwall of the fault zone, however, is a tough gouge seam from 1 to 4 inches wide, which is not crossed by ore seams. There is thus some suggestion of postmineral movement along a premineral fault that may have limited the ore deposition on the east.

---

47 Calkins, F. C., and Jones, E. L., jr., op. cit., p. 185.

The Alice, like the Morning and Gold Hunter lodes, is north of the Osburn fault but, unlike them, is very close to it. Three veins are recognized—the Alice, Mary J., and Meed Tunnel—the first two parallel to the fault and the third somewhat oblique. There are numerous slips, doubtless related to the Osburn fault, some of which are younger than the ore and offset it and some older whose gouges have served as effective barriers to the ore-depositing solutions. The ore follows ill-defined breccia zones, but individual ore bodies are of comparatively small extent; the largest is said to have yielded ore to the value of $70,000.

In the Carbonate Hill lode the mineralized rock consists of narrow and irregular quartz-siderite veinlets, many of which lie in the bedding planes, and of irregular replacement deposits along fissures, which contain most of the galena and sphalerite, attain a width of 3 or 4 feet in places, and afford bunches of ore distributed irregularly along the main drift for several hundred feet. Pyrite and locally chalcopyrite occur in the ore, and magnetite together with chlorite was observed at several places disseminated in the rock adjacent to the sulphide-bearing seams.

FORM AND DISTRIBUTION.

The ore bodies of the Morning and Gold Hunter mines are quite different in form and distribution, although both are tabular replacement deposits of large size. The Morning ore shoot, outlined in a general way by the stope sheet (fig. 4), forms an immense lens which is worked by continuous stopes as long as 1,600 feet and to a depth of about 3,200 feet below the surface without diminution in the extent of the ore body. The downward course of the ore shoot in general is nearly parallel to the line of dip of the vein. In the west end the lode splits about a great horse described in the two previous reports. On the 200-foot level the horse is 35 feet wide and separates two branches of the lode, each 10 feet wide. Above this level the horse is about 800 feet long, but below this level it shortens, and near the 1,050-foot level the branches unite, forming ore to a width of 40 feet. On the 1,450-foot level west another horse has been found which had not been fully delimited at the time of visit but seems to be about 300 feet long and on the east end causes the vein to branch, forming an angle of about 20° between the parts. The north branch bends south on the eighth floor above the level, at a point 240 feet from the east end of the horse, and reaches the plane of the south branch on the tenth floor; thence the line of junction pitches eastward to the main level. The extent of the horse below this level had not been determined at the time of visit.

The width of ore in different places in the mine ranges from a few inches to 40 feet or more, the prevalent width being perhaps 9 to 12

feet. Barren places and parts too low in grade to be worked at a
profit are of minor extent but have been found throughout the lode
and are fairly abundant near its ends.

The Hunter lode (Pl. XIII) is much more irregular in outline and
the ore bodies are distributed through a much shorter although a

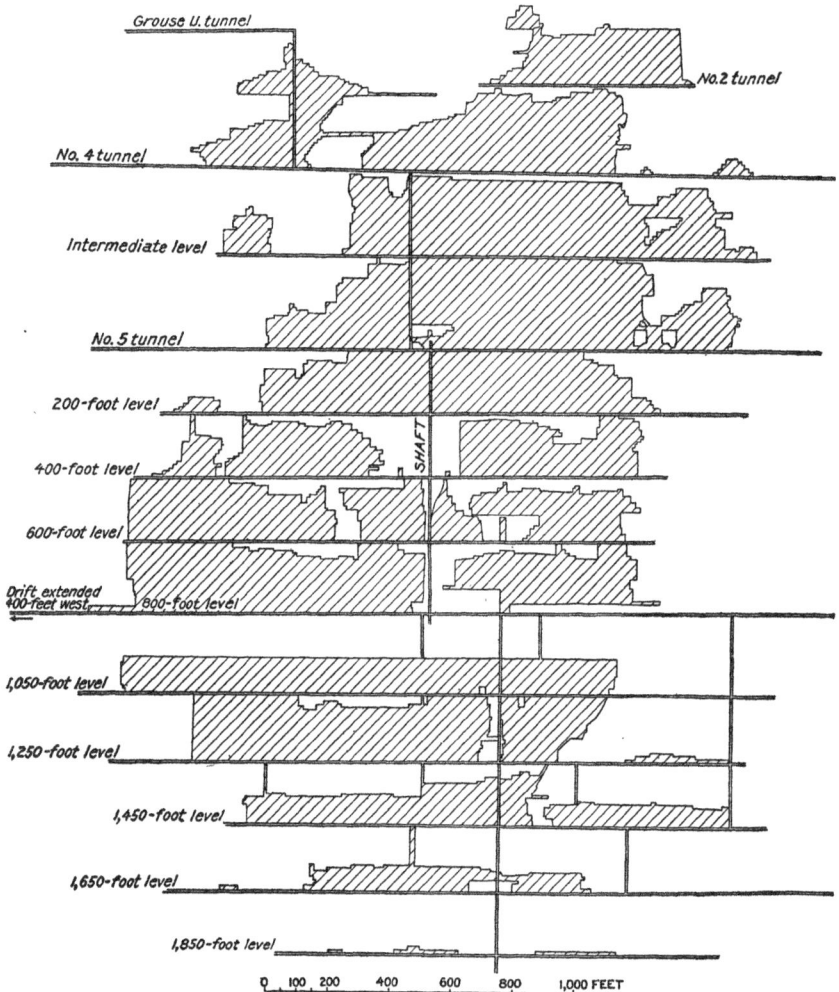

FIGURE 4.—Longitudinal section through the Morning mine, showing the principal stopes.

wider zone. In the upper levels the lode consisted of three principal
veins, which in depth merged with more or less distinct bodies distributed in a zone about 600 feet long and 100 feet wide. On the
600-foot level the three veins, known as the north, middle, and south
veins, are fairly distinct. The north vein is here 40 feet and the
south vein 60 feet from the middle one. On the 800-foot level only

two veins are recognized, and on the 1,200-foot level development at the time of visit had not defined the limits of the ore.

Within each of the veins on most levels there is more than one ore body, the outlines of which are exceedingly irregular in detail. Many bunches and patches of the ore minerals are too small to be mined at a profit.

THE ORES.

The Morning ore is valuable for lead, silver, and zinc and the Hunter ore for lead and silver. Siderite, accompanied by quartz and a little barite, is the dominant gangue mineral in both lodes aside from residual wall rock.

The Morning ore ranges in different parts of the lode from ore in which fine-grained massive sphalerite with a meshlike distribution of galena forms blocks 10 to 15 feet wide to ore of medium coarse grain in which galena predominates. In general the ore in the west end of the lode is fine grained, with sphalerite in excess of galena, and the ore in the east end is coarser, with argentiferous galena in excess. Barite is more abundant in the east end, and quartz, if any difference exists, attains its greatest development in the west end. Siderite, though varying markedly in relative amount from place to place, was not observed to exhibit a persistent variation from east to west. In most places the sulphides are distributed fairly uniformly across the lode, but locally there is a footwall and a hanging wall vein of somewhat different mineral composition. On the 1,200-foot level east of the shaft a hanging-wall vein is 4 feet wide, and about 65 per cent of this width is sulphide, in which galena is notably in excess of sphalerite; a footwall vein, 20 feet away, is 3 feet wide, and in it sphalerite and galena in nearly equal amounts make up perhaps 40 per cent of the mass. The intermediate ground is traversed by seams of the sulphides, which also occur in the walls, so that in mining a width of 20 to 30 feet is broken. Throughout the lode galena is the youngest mineral formed in noteworthy quantity. It has a meshlike distribution through sphalerite and the older siderite and quartz. Sphalerite is more characteristically in veins and bunches but locally forms a fretwork in the gangue minerals. On the whole quartz is younger than siderite, although much of it is so intergrown with the carbonate as to indicate contemporaneous deposition. Pyrite and locally pyrrhotite and magnetite accompanied both the siderite and the quartz. Calcite occurs in places and apparently formed contemporaneously with the iron carbonate. Barite in most places occurs as a fairly definite streak or as streaks from 1 inch to a foot or more wide. It envelops rounded masses of siderite and extends as hook-shaped protuberances into the siderite. In most places there is no evidence of difference in age between quartz and barite, but where one cuts the other the quartz is the younger.

Tetrahedrite occurs in some of the ore, but its age relations are not known. The galena-depositing solutions attacked the materials of the preexisting ledge matter in the following order, as shown by contact relations and enlargements along galena veinlets: (1) Siderite (and calcite), (2) quartz and sphalerite, (3) quartzite, (4) barite, which in most places was fully resistant.

The same paragenetic relations hold throughout the lode, whether lead or zinc predominates and regardless of textural types. In several places recurrence of conditions favorable to quartz deposition is shown by veinlets of quartz cutting galena, but these later veinlets nowhere attain quantitative importance. In one place on the 1,200-foot level east a 2-inch seam of clean sphalerite of normal dark-brown color traverses galena-siderite ore, a relation not observed in the Wardner district. The wider veinlets that extend off into the wall rock are predominantly sphalerite replaced by galena, and in the narrower ones consist of clean galena. In the main ledge also sphalerite, except as traversed by the minute galena seams, forms nearly clean masses locally 10 feet or more wide and continuous for many feet, but galena is nowhere found except in the small veinlets cutting other ore minerals and in bodies obviously due to enlargement or coalescence of the veinlets. It appears that the galena solutions traversed a mass most minutely shattered, whereas sphalerite was deposited in more widely spaced and very much larger openings.

The Gold Hunter ore differs from the ore of the Morning Lode principally in the relative abundance of the minerals. It is valuable for lead and silver, sphalerite though widely distributed occurring in small amounts. To 1 per cent of lead there is about 1 ounce of silver. Pyrite is more abundant here than in any of the other lead-silver mines of the county, except in the Blue Bird lodes of the Wardner area. Stibnite occurs as nests of acicular crystals in quartz. Barite forms dull white bands alternating with fine-grained galena and associated sulphides, which apparently were deposited along shearing planes in it. Locally oblique veinlets of the galena cut across barite bands. Quartz occurs in bunches, bands, and rounded kernels, some of which are embedded in barite. Siderite is clearly older than the galena and probably older than the quartz. The Gold Hunter differs from the Morning lode in that the openings formed during the general period of mineralization were almost entirely parallel to the vein walls, so that intersecting veinlets, the most satisfactory criteria of mineral sequence are but meagerly developed.

LODES ON CANYON CREEK.

SITUATION AND DEVELOPMENT.

The lodes on Canyon Creek northeast of Wallace include the Green Hill-Cleveland-Standard-Mammoth, the Tiger-Poorman, and the Hercules, on the north side of creek, and the Marsh, Hecla, and

Helena-Frisco, on the south side. They are distributed along the canyon from a point near the town of Gem to a point above Burke, a distance of 3 or 4 miles.

Most of the lodes are worked through tunnels and vertical underground shafts, but the Hecla and the Tiger-Poorman are opened from shafts starting at the surface. The Green Hill-Cleveland, the western continuation of the Standard-Mammoth lode, is worked through the Standard shaft, 2,250 feet deep. From the shaft a tunnel, used for transporting men and supplies, connects with the surface at Mace, and another, the old Mammoth No. 9, leads to the loading bins between Mace and Gem. Between the 1,200 and 2,000 foot levels the two mines are connected by drifts. The Hecla lode is developed to a depth of 1,712 feet below the collar of the shaft; the Helena-Frisco to 1,600 feet below the No. 6 or tunnel level; the Tiger-Poorman to 2,200 feet; and the Hercules through an 8,000-foot tunnel which taps the lode at a depth of 2,000 feet and from which a shaft had been sunk 500 feet farther at the time of visit. The Marsh mine is worked through a 900-foot shaft which opens from a tunnel from the Green Mountain claim. It was flooded at the time of visit.

Production of the principal mines of the Canyon Creek group.

Hecla.

Year.	Crude ore.	Concentrating ore.	Ratio of concentration.	Silver.	Lead.	Zinc.
	Tons.	*Tons.*	*Per cent.*	*Ounces.*	*Pounds.*	*Pounds.*
1898–1905				2,042,726	67,418,922	
1906	2,596	100,631	7.16	537,437	17,778,942	
1907	7,062	95,431	8.52	538,390	18,764,978	
1908	4,211	54,991	8.44	289,885	10,876,250	
1909	7,757	100,748	9.51	481,475	18,348,112	
1910	7,247	98,993	9.35	492,420	17,562,222	
1911	6,059	102,865	7.05	493,880	18,865,146	
1912	7,137	100,760	7.76	491,745	19,137,960	
1913	6,622	99,305	7.46	507,236	18,832,534	
1914	7,037	90,851	6.98	509,200	18,957,823	
1915	11,015	112,646	7.41	692,444	24,917,867	
1916	16,321	184,209	7.52	1,195,841	40,217,573	
	83,064	1,141,430		8,272,679	291,678,329	

Tiger-Poorman.[a]

Year.	Crude ore.	Concentrating ore.	Ratio of concentration.	Silver.	Lead.	Zinc.
1904 [b]		129,868	9.9	343,160	13,900,000	
1905	22	132,381	10.8	317,726	12,832,600	
1906		116,667	10.9	287,932	11,169,800	
1907		95,200	10.4	235,548	8,657,000	
1908		14,960	6.4	45,418	2,110,000	
1909	434	11,209	7.6	40,179	1,578,000	
1910						
1911						
1912	118			3,107	108,600	
1913	77			1,907	62,200	
1914 [b]	100			3,472	92,400	[c] 23,656
1915 [b]	99			1,910	60,600	
1916 [b]	58			1,481	45,282	
	908	500,285		1,281,840	50,616,482	23,656

[a] The gross production prior to the end of 1905, as given by Ransome, was 811,794 tons, containing 89,199,398 pounds of lead and 286,424 ounces of silver.

[b] For the period prior to 1914 the figures are for the 12 months ending Aug. 31; for 1914, the 16 months ending Dec. 31; for 1915, the calendar year; and for 1916, the 5 months ending May 31.

[c] Lessee's ore only.

Production of the principal mines of the Canyon Creek group—Continued.

Standard-Mammoth.[d]

Year.	Crude ore.	Concentrating ore.	Ratio of concentration.	Silver.	Lead.	Zinc.
	Tons.	*Tons.*	*Per cent.*	*Ounces.*	*Pounds.*	*Pounds.*
1904 e	754	373,557	7.9	1,891,740	45,877,400	
1905	4,367	357,865	8.8	1,747,333	44,146,000	
1906	5,087	332,911	5.8	2,632,209	58,854,200	
1907	2,917	293,533	5.2	2,420,768	50,599,600	
1908	8,275	278,025	6.4	2,021,660	45,393,000	
1909	13,182	264,218	7.6	1,775,541	42,502,600	
1910	9,009	201,891	7.1	1,348,642	33,537,520	
1911	13,914	209,236	7.9	1,630,922	36,725,600	
1912	10,910	194,319	9.9	1,159,196	27,333,600	
1913	1,161	22,728	8.3	174,931	3,565,200	
1914 e	26			1,572	f 25,000	
	69,602	2,528,283		16,804,514	388,559,720	

d Prior to absorption by the Federal Co. the Mammoth paid $2,945,000 and the Standard more than $1,500,000 in dividends. Ransome gives the production from the consolidation by the Federal Co. to the end of 1905 as 850,102 tons of ore containing 109,697,511 pounds of lead and 4,473,860 ounces of silver.

e For the period prior to 1914 the figures are for the 12 months ending Aug. 31; for 1914, the 16 months ending Dec. 31.

f Subsequent to 1914 included with Greenhill-Cleveland production.

Greenhill-Cleveland.

Nov.–Dec., 1912	2,831	36,313	7.8	365,669	6,060,310	
1913	13,723	179,752	6.7	1,859,588	29,341,427	1,083,095
1914	13,013	177,698	6.4	2,085,909	33,327,453	5,701,051
1915	9,308	161,013	7.8	1,363,550	23,022,071	g 3,800,581
Jan.–Mar., 1916	1,962	31,022	7.5	257,000	4,666,950	357,060
	40,837	585,798		5,931,716	96,418,211	10,941,787

g Subsequent to 1914 includes Standard-Mammoth production.

Helena-Frisco.

1897 (5 months) to 1906		1,042,640		3,896,801	108,728,324	
1907		55,846		196,990	6,758,658	2,244,655
May–Dec., 1915 h		43,513	8.3	36,411	1,620,940	2,479,174
Jan.–May, 1916		43,850	8.5	15,381	1,018,918	2,670,650
		1,185,849		4,135,583	118,126,840	7,394,479

h From 1908 to May, 1915, the upper levels were worked intermittently on a small scale by lessees. Production for this period is not available.

Hercules Mining Co.

Year. i	Shipping product (tons).	Gross value.	Year.	Shipping product (tons).	Gross value.
1907	20,466	$1,615,683	1913	40,816	$2,846,758
1908	19,445	1,195,878	1914	60,559	4,147,609
1909	17,950	1,059,036	1915	49,441	3,090,174
1910	25,765	1,249,081			
1911	31,399	1,922,486		300,036	19,300,750
1912	34,195	2,174,045			

i Production prior to 1907 as given by Ransome, 62,648,676 pounds of lead and 3,920,838 ounces of silver. Later production from statements filed with county assessor and compiled by the Wallace Miner, issues of Dec. 30, 1915, and May 11, 1916.

GEOLOGIC RELATIONS.

The lodes along Canyon Creek are all inclosed in the Burke formation, which forms both walls of the canyon for about 5 miles east of the monzonite mass at Gem, or in transition beds between the Burke and the Prichard. The formation in general strikes north and forms an anticline whose axis is in the vicinity of Mace. Thus in the Hecla mine the beds dip eastward at angles of 70° to 85°, and in the Standard-Mammoth and Green Hill-Cleveland they dip westward at comparable angles. Farther west the beds are upturned and surface exposures show a steep eastward dip. This general relationship is of particular interest, as the lodes have a course almost at right angles to the axes of folding, and if the type or degree of mineralization is controlled by stratigraphic position persistent east-west variations should appear along each level in lodes like the Standard-Mammoth, which has been stoped continuously for more than 2,000 feet. The west end of this lode on any particular level is perhaps 1,800 feet higher stratigraphically than the east end. The deeper levels are in the transition beds between the Prichard and the Burke formations, where, according to Hershey,[48] lead deposits are supposed to give out. Below the 1,250-foot level the ore body pitches westward on the east end but the west end is nearly vertical (fig. 5).

The Tiger-Poorman lode crosses beds which dip east at an average angle of 65°. The ore shoot separated into three parts on the 1,700-foot level, two of which continued to the 2,000-foot level and one to the 2,200-foot level, below which ore was not found (fig. 6). The ore body as a whole had nearly vertical sides above the 1,800-foot level and extended to the greatest depth below the middle prong. The inclosing beds on the lower levels are said to be Prichard slate, so that any close stratigraphic control of ore deposition should certainly be expressed in a marked eastward pitch of the ore, at least within the transition zone between the Burke and Prichard beds.

The Hecla ore bodies show similar independence of the type of the inclosing formation, although this mine is not yet deep enough to show critical relations. The Helena-Frisco lode (fig. 7), comprising three shoots separated by faults and known as the Gem, Frisco, and Black Bear, is inclosed in the Burke formation near the monzonite contact. The beds here predominantly dip east at steep angles. The middle shoot has been found productive to the 2,200-foot level; the others to about the 1,000-foot level. Admission to the Hercules mine was refused to the writers, but this mine is understood to contain three ore shoots along a lode which crosses steeply dipping beds. Two of the shoots apex at a depth of 1,000 to 1,500 feet and unite with the main shoot to form an ore body 1,800 feet long on the No. 5 tunnel level. Development has extended for 500 feet below this level, and there is said to be some evidence of further extension of ore to the east, although lower levels have not been so promising.

Lamprophyre dikes were observed in the Hecla, Helena-Frisco, Standard-Mammoth, and Greenhill-Cleveland mines. In the Helena-

---

[48] Hershey, O. H., Origin and distribution of ore in the Coeur d'Alene (private publication), p. 23, 1916.

Frisco a basic dike 3 feet wide, with exceptionally sharp contacts, cuts across the lode in a small stope west of the shaft on the 1,600-

FIGURE 5.—Longitudinal section through the Standard-Mammoth and Greenhill-Cleveland mines, showing recent stopes.

foot level. The vein here consists of innumerable seams and bands of sphalerite in a light-gray quartzite. The dike consists of scattered phenocrysts of augite and small grains of magnetite in an aggregate of interlocking hornblende and plagioclase crystals. It is classified as camptonite. Two dikes not in contact with ore were observed in the Standard-Mammoth mine east of the shaft. One of them crosses a flat fault of reverse throw that puts slate above quartzite with discordance in dip. Another dike, 5 feet wide, appears in opposite sides of the stope above the 2,050-foot level and west of the shaft. As stoping has here been completed, its age relation to the ore is not known. Microscopic examination shows the rock to be

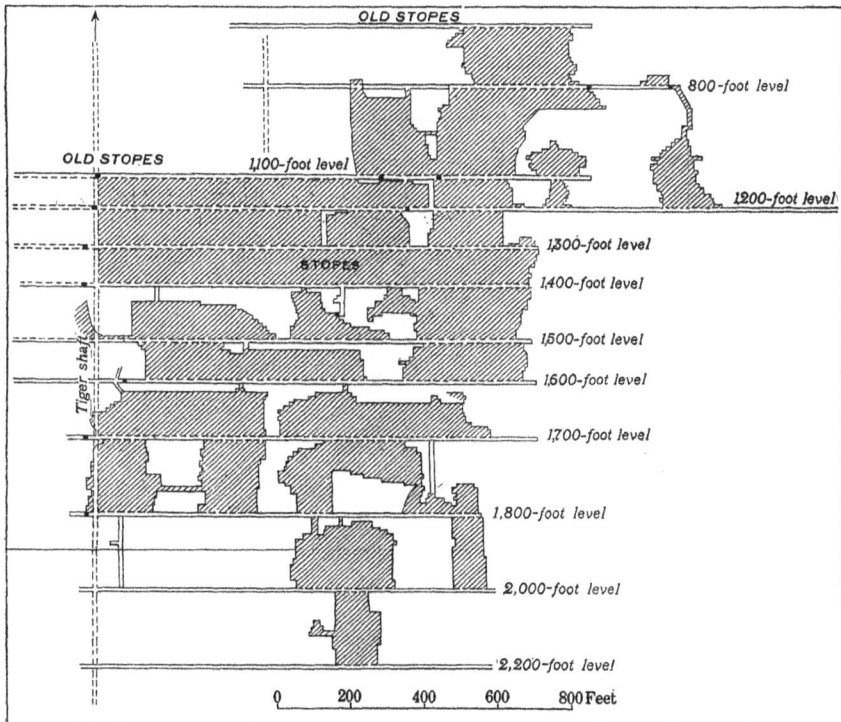

FIGURE 6.—Longitudinal section of the Tiger-Poorman mine.

camptonite, similar to that in the Helena-Frisco mine, but with the augite greatly altered to minutely crystalline aggregates, largely sericite, serpentine, and amphibole.

A dike averaging perhaps 2 feet in width follows the Hecla lode more or less continuously, the ore occurring on one or both sides of the dike. Locally the ore seems to replace the dike material, but thin sections from such places show the contact to be an intrusive one, with marginal chilling of the igneous material and minute dikelets of it extending into the ore. The dike is clearly younger than the

109496—23——6

vein, as pointed out by Ransome.[49] His description of materials from the upper levels, however, indicates a rock notably different from that collected by the writers. He describes "a holocrystalline aggregate of abundant phenocrysts of hornblende in a groundmass of the same mineral, with calcic plagioclase in minute laths." The four specimens studied by the writers, two of which are of exceptionally fresh material from the 900 and 1,600 foot levels, contain biotite pheno-

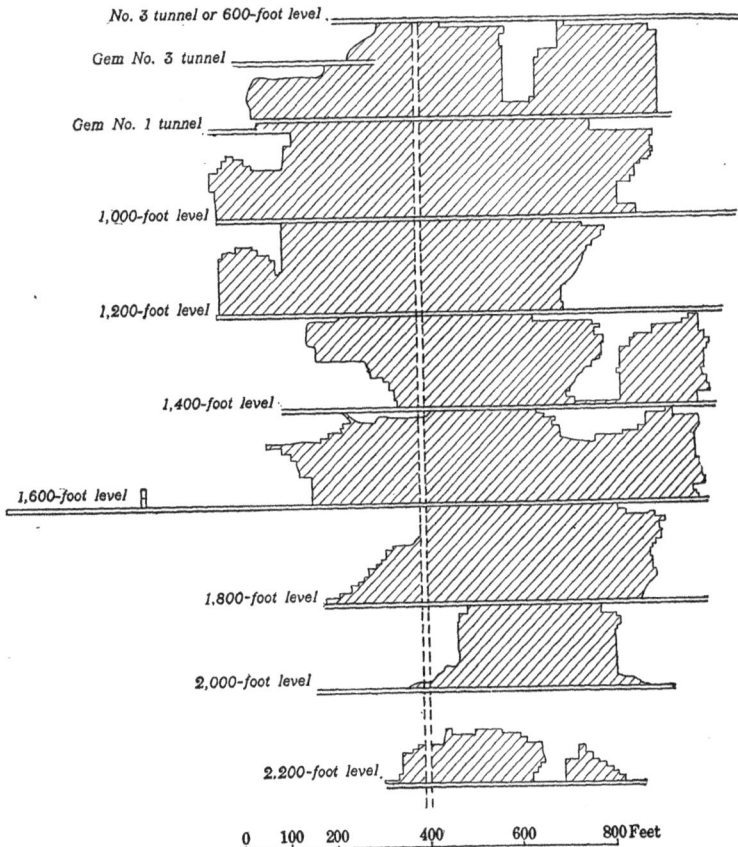

Figure 7.—Section of the Frisco mine, showing principal stopes below the No. 3 level.

crysts in an orthoclase-biotite groundmass. A little augite may be represented by aggregates of secondary calcite, serpentine, and chlorite. Hornblende is rare, and no feldspar with higher index than Canada balsam was observed. This rock is minette, whereas the specimens studied by Ransome are odinite, "hornblende being too abundant for kersantite." As the dike has been continuous down through the lode, it seems necessary to conclude that it changes from a hornblende-plagioclase rock above to a biotite-orthoclase rock below.

[49] U. S. Geol. Survey Prof. Paper 62, p. 176, 1908.

Faulting of the Canyon Creek lodes is not a prominent feature of their geologic relations. The Standard-Mammoth lode terminates against a fault above the 1,250-foot level, which Ransome concluded to be premineral, a conclusion borne out in recent years by fruitless search for a continuation of the ore east of it. On the west end, well over in the Greenhill-Cleveland property, the lode on the 1,450-foot level terminates against a north-south fault which dips 80° E. The lode, only meagerly mineralized at this point, leads to the fault on a course N. 70° W. and runs north for a short distance along the fault plane. The fact that the vein quartz is not brecciated at the turning point strongly suggests that the fault is older than the mineralization. This age relation is further suggested by exploration beyond the fault, where two crosscuts to the south, 200 and 300 feet long, and a 2,000-foot crosscut to the north have failed to reveal a continuation of the lode.

The Helena-Frisco lode consists of three parts, locally known as the Gem, Frisco, and Black Bear veins, separated by faults or cross fissures, which cause offsets to the north on the east side. Ransome [50] raised the question whether this might not be a result of premineral faulting, basing his query on a mineralogic difference between the veins. Nothing was observed during the recent examination which throws light on this problem, although the operators agree in considering the three veins as fault segments of one nearly east-west lode. The Tiger-Poorman lode ended sharply on levels 8 and 12 against the O'Neil Gulch fault, but on other levels it failed to reach the fault.

FORM AND DISTRIBUTION.

The lodes of the Canyon Creek group are tabular replacement deposits which strike north of west except the Hecla, which strikes northwest, and dip southwest except the Standard-Mammoth-Greenhill-Cleveland, which dips very steeply northeast. The ore shoots are distributed along the lodes without apparent relation to curvature or cross fissuring. Between them the lode is marked by more or less shearing and in many places by interrupted seams of quartz or siderite. Locally on their margins, as in part of the Greenhill-Cleveland mine, the productive shoots merge into crushed rock containing much pyrite.

The Tiger-Poorman ore body separates downward into three limbs and the Standard into two limbs, as shown in figures 5 and 6. In the Hecla and Hercules lodes a converse relation occurs, three shoots being recognized on the lower levels, whereas only one was found above. The two additional shoots cut on the lower level of the Hercules mine are said to apex 1,500 feet below the surface.[51] In

[50] Op. cit., p. 180.
[51] Bell, R. N., oral communication.

the Hecla mine the upward limit of the middle shoot is between the 600 and 900 foot levels. The east shoot has been explored on the 300 and 900 foot levels. Above the 300-foot level it is developed by a raise for 600 feet, but below the 900-foot level drifts have not been extended to it.

In much of the Tiger lode, according to Ransome,[52]

the ore occurs in a series of overlapping or imbricated, nearly vertical fissure zones. * * * Three of these are usually recognizable and are known as No. 2, No. 3, and No. 4 ore shoots. Each successive shoot is usually found by crosscutting to the right when the ore in the shoot worked shows signs of pinching, the crosscut in many cases following a linking fissure. On some levels, however, branches from one ore shoot extend obliquely to the other and no crosscutting is necessary.

The Hecla lode contains three shoots distributed along a nearly vertical zone of intense shearing which strikes N. 40° W. The northwest ore shoot, upon which the early work was done, has been worked out from the surface down to the 900-foot level and developed to the 1,600-foot level. It pitches east at a steep angle, approaching within 40 feet of the middle shoot, a westward-pitching body, on the 1,200-foot level. Here there is some suggestion that the shoots overlap, the middle one lying southwest of the other. Development on the two lower levels, however, had not demonstrated such a relation at the time of visit. On the 1,200-foot level the combined shoots have a stope length of more than 1,300 feet and an average width of perhaps 6 or 8 feet, although ore seams which extend into the walls obliquely or follow shearing planes parallel to the main ore body are sufficiently abundant to warrant mining to a width of 10 or 12 feet in most places and locally to a width of 30 feet or more. In most exposures perhaps one-half of the ore in the middle part of the lode is of shipping grade owing to the presence of seams of fine-grained galena which attain 2 or 3 feet in width. Adjacent to this ore narrow seams occur largely along the innumerable shearing planes in the quartzite. Very little ore has been extracted from the middle shoot, although it has been rather fully developed between the 900 and 1,600 foot levels, and no ore has been taken from the east shoot except in exploring it on the 300 and 900 foot levels. On the former it is 850 feet long. In no place do any of the numerous ore seams that cut across the sheeted quartzite of the lode traverse the minette dike, and many of them may be seen to terminate abruptly against it. Ore occurs on one or both sides of the dike, which in most places must be broken in mining.

The Standard-Mammoth lode, including its westward continuation in the Greenhill-Cleveland mine, differs but little from the lodes already described. It contains one great ore body nearly 2,000 feet

---

[52] Op. cit., p. 173.

long on most levels, which has been worked continuously to a vertical depth of more than 3,500 feet and ranged from 5 to 15 feet in width. Below the 1,650-foot level two limbs, 225 and 450 feet long on the 2,050-foot level, are recognized. The ore largely replaces sheared quartzite, although crevice and fissure filling, as in most of the Canyon Creek group, played a noteworthy part in its development. Ransome[53] observes that "a very characteristic feature of the lode is the presence of numerous nearly horizontal stringers of quartz which are later than the mass of the ore." This feature is almost entirely absent in the lower levels now accessible. In one stope above the 1,050-foot level quartz seams later than the other minerals were noted, but here most of the quartz is contemporaneous with or older than the sulphides; elsewhere in stopes now open quartz represents an early stage of mineralization.

The Helena-Frisco lode as here considered embraces three veins, the Black Bear, Frisco, and Gem, separated by faults and probably segments of a once continuous fissure. Recent studies throw no light on the problem of whether the faulting preceded or followed mineralization. The Black Bear Fraction vein may be a fourth segment separated from the Black Bear by an oblique fault, but as present exposures only suggest the possibility of such a relationship, it is here considered a part of the Black Bear vein.

The lode contains five ore shoots, two each on the Black Bear and Gem veins and one on the Frisco. On the west the larger Gem ore body ended against the monzonite and was worked nearly to the 1,200-foot Frisco level. It attained a maximum length of about 800 feet but below the 600-foot level rapidly shortened downward. East of it a shoot averaging about 125 feet in length was worked nearly to the Frisco 1,000-foot level. The principal shoot of the lode is the Frisco, and it is almost coextensive with the Frisco segment of the fissure. It was a body from 10 to 12 feet wide and about 800 feet in average length and was worked to the 2,200-foot level. The Black Bear shoot was more irregular in outline, although better-defined walls gave it a more veinlike aspect. It filled a fissure from 2 to 3 feet wide somewhat interruptedly to a depth of 700 feet. On the Black Bear Fraction, a property under different ownership, is a shoot of ore only partly blocked out which averages between 5 and 6 feet in width and is at least 300 feet long where explored from a 2,500-foot tunnel at a depth of about 2,000 feet. The ore shoot is developed by two raises from this level, each reaching 200 feet, another raise 70 feet up, an upper tunnel 600 feet long at a level 1,180 feet higher, and a few surface cuts. The vein, which strikes N. 70° W. and dips 78° SW., consists of numerous seams and veinlets of sphalerite, galena, and chalcopyrite in a quartz-siderite gangue.

[53] Op. cit., p. 178.

The ore contains 6 to 8 per cent of lead, 10 to 15 per cent of zinc, and between 2 and 3 ounces of silver to the ton.

THE ORES.

The lead ores of the Canyon Creek group are unusually rich in silver among the Coeur d'Alene deposits, although the ratio of silver to lead shows a wide range in different deposits and in different parts of the same deposit. The contents as set forth in the following table are derived principally from statistics of production and consequently do not reveal local variation within individual deposits.

Relation of silver and lead in Canyon Creek lodes.

	Silver (ounces to the ton).	Lead (per cent).
Hecla	2.6	4.3
Hercules	1	1
Tiger-Poorman	1	1
Standard-Mammoth	1.6	1.9
Greenhill-Cleveland	5.9	4.8
Helena-Frisco a	{ 3.9	5.4
	4	13
Black Bear Fraction	1	5

a The upper figures are for the period 1897 to 1903 and the lower for the years 1915–16.

It is apparent from the above table that the west end or the Greenhill-Cleveland part of the Standard-Mammoth lode contains much more silver in proportion to lead than the east end, also that the Black Bear Fraction or eastern extension of the Helena-Frisco lode is lower in silver than the main shoot in the Frisco mine. The variation in the amount of zinc in different members of the group is such that the Helena-Frisco as now worked produces almost twice as much zinc as lead. The Tiger-Poorman, Standard-Mammoth, and Hercules have produced little or no zinc. Recently a zinc vein oblique to the lead vein has been found in the Hecla. In the Greenhill-Cleveland the ratio of zinc to lead is about 1 to 9, and in the Black Bear Fraction 2 to 1. There is a general belief among the operators of mines along Canyon Creek that sphalerite increases with depth, although examination at any one stage in development affords little basis for such a conclusion, owing to a marked difference in the relative amount of sphalerite in different parts of the same stope.

Chalcopyrite, pyrite, and pyrrhotite are present in most of the lodes, and magnetite occurs in the Helena-Frisco. Chalcopyrite attains its maximum development in the Black Bear Fraction, where in raise No. 1 it is fairly abundant in intimate association with galena and sphalerite, which replace quartz and quartzite. Much of the Helena-Frisco ore contains visible chalcopyrite, though nowhere in economic quantity. In the Standard-Mammoth lode chalcopyrite occurs locally in small amounts; in the other lodes it is exceptional. Pyrite is nearly everywhere present but in widely different amounts. It is more widely disseminated in the quartzite than any of the other sulphides and is almost universally present in the lodes, being particularly abundant in the peripheral parts of ore bodies. Above the 2,050-foot level on the Mammoth claim pyrite occurs almost to the total exclusion of galena and sphalerite in many places, but on the 2,250-foot level on the same shoot it is not a conspicuous element of the ore. Here the ore is principally sphalerite in a siderite gangue and occurs as innumerable narrow seams in firm quartzite. Ransome[54] reports pyrrhotite "in considerable quantities in the lower levels of the Tiger-Poorman and Standard-Mammoth mines" and as "a minor constituent of the Helena-Frisco ore." The Tiger-Poorman, now flooded, has been worked little since his visit, but the Standard-Mammoth has been continued 1,200 feet deeper, and on these lower levels, now alone accessible, pyrrhotite was suspected in only a few places, and none of it is present in the material collected for detailed examination. Also in the Helena-Frisco lode pyrrhotite is rare in stopes now being worked, but magnetite, not mentioned by Ransome, is locally conspicuous and a magnet reveals it in crushed portions of much ore that is apparently devoid of it. It occurs intimately mixed with steel galena in seams that traverse dark-brown sphalerite. In one specimen nests of fibrous actinolite occur in magnetite-galena areas.

Siderite occurs in conspicuous amounts in the Standard-Mammoth lode and greatly exceeds quartz on the deepest levels. In the Hecla lode, however, quartz is the dominant gangue mineral in the lower part of the mine, and siderite only locally exceeds it in amount on the upper levels. Here calcite is developed to a minor extent. No carbonate was observed in the west end of the Helena-Frisco lode, but in the Black Bear Fraction siderite is locally as abundant as quartz.

The paragenetic relations of the minerals of the Canyon Creek lodes are not as clear as those in the other lead-silver deposits of Shoshone County. In many of the lodes there is abundant evidence of the recurrence of conditions favorable to the deposition of a particular mineral or group of minerals. Quartz in general is younger than siderite and in part contemporaneous with it, but in the Hecla and Standard lodes quartz seams traverse the lode and extend into the wall rock. In the Helena-Frisco ore the quartz is the oldest mineral now in the lode. It is replaced by sphalerite, which is traversed by seams of galena, and in the west end on the 1,800-foot level by seams of galena and magnetite intimately intergrown and locally

[54] Op. cit., p. 109.

containing felted aggregates of actinolite. In the side of the drift near by a veinlet of vesuvianite was observed cutting monzonite, through which pyrite and a little sphalerite are disseminated. Magnetite seen elsewhere in the lode is older than sphalerite and is not associated with lime silicates. The relations are considered as proof of a later pulse of mineralization characterized by higher temperature and more complex solutions than that which initiated deposition along the lode and formed most of it.

The position of chalcopyrite and tetrahedrite in the sequence of mineral deposition is not definitely known, although the former is clearly older than galena and younger than quartz. Pyrite may have had a long period of formation, but most of it was developed about contemporaneously with siderite. Throughout the group galena is younger than sphalerite and, except for a second generation of quartz developed in places, is clearly the last mineral to form in the lodes. This relationship is clearly apparent in the stopes, where seams and bands of galena traverse the lode in a most intricate manner. In many polished specimens the evidence is conflicting owing to postmineral movement along the lodes which has modified contacts in minute detail.

LODES ON NINEMILE CREEK.

The most valuable mines on Ninemile Creek and the head of Beaver Creek, immediately to the north, are primarily zinc producers, and their lodes are described elsewhere (pp. 90–105). There are, however, several which have yielded more lead than zinc. These are the Tamarack and Custer, Sunset, Idora, Tuscumbia, and several others not now active. Formerly the Callahan, Rex (Sixteen-to-One), and Success (Granite) mines were worked for lead, but now zinc is the chief metal of their ores.

The deposits, except the California, which is inclosed in the Revett quartzite, occur in the Prichard and Burke formations. In general they are steeply dipping lodes which strike north of west.

The Tamarack and Custer mine, controlled by the Day interests, is an old mine reopened in 1913 and active since that time except for short periods of idleness. During the three years prior to 1916 the mine produced ore to the gross value of $2,600,921, principally from lead. As access to the mine was not granted the only information available is that recorded by Ransome.[55] Prior to 1904 it produced ore of a total gross value of $600,000. The lode occurs in Burke quartzite high up on the west slope of Custer Peak. It is a well-defined fissure containing galena and a little chalcopyrite and pyrite in a quartz gangue.

The Sunset mine, near the summit of Sunset Peak, develops a lode 2 to 10 feet wide which extends from the monzonite on the east

---

[55] Op. cit., pp. 182, 183.

across a patch of Burke rocks into underlying Prichard on the west. A tunnel starts on the south slope, runs mostly through monzonite 1,800 feet to the vein, and thence follows it westward 1,260 feet to the surface northwest of the peak. At the northwest portal a vertical shaft extends to a level 600 feet lower. All the ore is in sedimentary rock, although a claim to the east contains seams of galena and pyrite in the monzonite. On the 400-foot level from the shaft two lodes are developed, both with poorly defined walls and only a meager sprinkling of sulphides. The better material consists of an intimate mixture of sphalerite, galena, pyrrhotite, chalcopyrite, and magnetite, together with quartz as seams and bunches in the quartzite. Garnet appears in much of the material on the tunnel dump, and a greenish pyroxene is present in some of it. The silicates, together with the magnetite and pyrrhotite, are in most intimate intergrowth with sphalerite and in some specimens with galena and chalcopyrite, clearly indicating conditions of high temperature and intense pressure at the time of deposition. Mineralogically the ore is of contact-metamorphic type, but as it follows a fissure across beds of diverse composition and extends almost at right angles to the igneous contact, it is classed as a high-temperature vein rather than as a contact deposit. It is particularly noteworthy that the monzonite is pyritized and sericitized and contains lead and zinc in seams at a point near by. This suggests that the Sunset lode if prospected up to the monzonite contact might reasonably be expected to extend into it. Certainly the mineralization, at least in part, occurred after the marginal solidification of the monzonite.

Northwest of the Sunset mine, on the headwaters of Carbon Creek, are several lodes on the Idora and Tuscumbia groups of claims. Two lodes in the upper of two tunnels on the Idora group have been worked intermittently for a number of years and afforded a small production. They are inclosed in gray and blue Prichard slate, which dips east at low angles. The deposits are of the normal metasomatic fissure type and from 3 to 4 feet in width. The maximum length of stopes is about 300 feet on ore of medium to coarse grain consisting of galena, sphalerite, pyrite, and sparse chalcopyrite in a quartz-siderite gangue. Pyrite is abundant locally. The ore as mined contains from 5 to 6 per cent each of lead and zinc and a little less than 3 ounces of silver to the ton.

The California mine, situated in the canyon of Black Cloud Creek, a tributary of Ninemile Creek from the west at a point about 2 miles from its mouth, was not producing in 1916 and has been worked only intermittently in recent years. All told it has produced perhaps $250,000. It occurs in a down-faulted block of the Revett formation, whereas the other deposits in the Ninemile area are all in the Prichard, except the Sunset, which is in the lower part of the Burke. The ore occurs in a lode with poorly defined walls and consists of galena, pyrite, and quartz. Much of the quartz occurs in stringers that extend into the wall rock.

OTHER LEAD-SILVER LODES OF SHOSHONE COUNTY.

Lead-silver deposits well removed from any of the three groups described above are widely distributed in Shoshone County. They have not contributed much to the production of the county, although in an area of less intense activity several of them would attract considerable attention. As outliers of highly productive deposits they are of particular interest in defining the broad area in which lead deposition took place and in which there is ground for reasonable hope that new discoveries will be made.

PINE CREEK.

The Hypotheek mine, near the head of French Gulch west of Pine Creek, produced about $90,000 between June, 1916, and January 1, 1917, from concentrates averaging 50.65 per cent of lead and 6.23 ounces of silver and 0.125 ounce of gold to the ton, and crude ore containing 75 per cent of lead and 8.4 ounces of silver and 0.0036 ounce of gold to the ton. The lode was known as early as 1887, but yielded little until the recent discovery of a lead shoot in its west end beyond a transverse fault. It is developed by a shaft 1,100 feet deep. It strikes N. 70° W. and dips 70° SW. and traverses slate and quartzite of Prichard age. The lode is characteristically a fissure filling, although in many places it comprises a network of quartz veinlets in slate. The principal ore shoot is west of a fault that strikes N. 40° W. and dips steeply southwest. It is 400 feet long and from 5 to 6 feet in average width above the 900-foot level, but on the 1,100-foot level it splits into limbs, of which the longest is 75 feet in length. East of the fault and in its footwall the vein consists of quartz and carbonate in which are small and sporadic bunches of chalcopyrite, pyrite, galena, tetrahedrite, and arsenopyrite. The west shoot is oxidized down to the bottom of the present workings, which extend to a level 700 feet lower than the bed of Coeur d'Alene River. The ore is a mass of iron and manganese oxides with reticulating crusts of quartz and cerusite and scattered nodules of galena coated with cerusite crystals along with rare massicot and malachite. It averages about 9 per cent of lead. Another vein 175 feet north of the main one is explored for several hundred feet on the 1,100-foot level. It contains a shoot 100 feet long of galena, pyrite, chalcopyrite, and tetrahedrite in a gangue of quartz, calcite, and siderite.

Several other deposits in the Pine Creek area contain notable amounts of lead and silver, and it is possible that further development will show some of them to be more valuable for lead than for zinc. At present, however, they are considered zinc deposits and are described as such (pp. 105–109). All are inclosed in the Prichard formation. For purposes of cross reference mention should be made of the Douglas, in which, to judge from data now available, the lode contains about 6 per cent of lead and 20 per cent of zinc; the Constitution, whose ore as shipped contains 10 per cent of lead and 27 per cent of zinc; the Highland Surprise, with ore containing 3 per cent of lead and 16 per cent of zinc; the Northern Light, 5 per cent of lead and 8 per cent of zinc; the Little Pittsburg, 5 per cent of lead and 9 per cent of zinc; and the Denver, where assays show 31 per cent of lead, 20 per cent of zinc, and 19 ounces of silver to the ton.

MURRAY.

The Jack Waite mine is near the head of Eagle Creek, about 6 miles north of Murray. It is inclosed in Prichard beds at horizons probably between 3,000 and 4,500 feet below the base of the Burke quartzite. The vein strikes N. 55° W. and dips 52° SW., crossing the bedding of the inclosing slates and quartzites, which dip 20°–40° NE. The deposit is a quartz-filled fissure that ranges from a fraction of an inch to 10 feet in width and averages perhaps 4 feet. A gouge seam marks the fissure, and in many places quartz has been deposited along both sides of it, leaving the gouge almost entirely unreplaced. The developments consist of two principal tunnels, of which the lower one reaches the vein by a crosscut 943 feet long and thence explores it toward the southeast for 1,500 feet. Two principal quartz shoots are revealed by this drift, one 250 feet long and another about 300 feet long. Galena replaces the quartz sparsely in the first shoot, but in the second it is locally abundant and in one place is being stoped for a distance of 70 feet along the level. Beyond this are two other shoots, one 18 and the other 30 feet in length. An upper tunnel 365 feet higher enters on the vein and follows it for 1,500 feet, opening the two quartz shoots seen on the lower level. The west shoot contains ore for 75 feet and the east shoot for about 50 feet. On the surface the vein crops out in the bottom of a narrow gulch up to an elevation 500 feet higher than the upper tunnel. Here the ore consists of quartz spattered with cerusite and locally pyromorphite, together with much unoxidized galena. Beneath the surface the ore consists of galena and a little sphalerite, which clearly replace the quartz. The ore as shipped, after being hand-picked, contains about 60 per cent of lead, 3 per cent of zinc, and 6 to 10 ounces in silver to the ton. Few shipments had been made at the time of visit, but active development work was in progress and shipping has been started since.

Several prospects in the vicinity of the Jack Waite are only slightly developed and were not visited. In Oregon Gulch a strong lode occurs along the Murray Peak fault. It is about 6 feet wide, as shown in several open cuts and shallow pits, and consists of quartz containing much pyrite and pyrrhotite and a little galena and sphalerite. A similar ledge occurring in alinement with the Murray Peake fault and possibly a continuation of the one just described crosses Prichard Creek a short distance east of Murray. It is known as the Gold Back ledge. The developments consist of an open cut and a tunnel 700 feet long. The open cut exposes a wide zone, through which quartz and associated sulphides are distributed in bands separated by slate and so spaced that perhaps 60 per cent of a zone 10 feet wide is quartz. The sulphides include pyrite, sphalerite, galena, pyrrhotite, and chalcopyrite, all in small amounts and irregularly distributed. Assays yield about 80 cents in gold to the ton and a little silver. The inclosing rock is Prichard slate.

PRICHARD CREEK.

The Monarch lode, 5 miles southeast of Murray, on the south side of Prichard Creek, has been explored to a depth of 1,465 feet below the discovery shaft. The inclosing rocks are the quartzite and slate which form the transition beds between the Burke and Prichard formations. The developments include a crosscut tunnel 3,292 feet long, which reaches what is probably the Monarch ledge at a distance of 2,371 feet from the portal and a heavy fault gouge at 2,669 feet. Here a drift 1,000 feet eastward follows the fault, which dips 60°–75° S. and is in about the position which the ledge would occupy if unfaulted. A crosscut to the north in the east end of the mine reveals the ledge 200 feet in the footwall of the fault. Other tunnels are known as No. 1 and No. 2, respectively 1,331 and 1,112 feet above the 1,400-foot or lowest level. Several intermediate levels are connected with the main ones by winzes and raises. Ore was stoped from the surface to a depth of 1,000 feet and to a height of about 200 feet above the 1,400-foot level. The fault mentioned above must pass between these points and probably accounts for a marked lack of alinement of the upper and lower sets of stopes, but as this part of the mine was not safely accessible at the time of visit no detailed observations were made. As shown by stope maps the ore body pitched about 60° E. and was stoped for an average length of about 175 feet. The lode strikes N. 50° W. and stands nearly vertical above the fault, but below the fault it dips 80° NE. The ore as seen in a few places on the 1,100 foot level consists of galena, some sphalerite, and a little pyrite in a quartz gangue. In places there is much disseminated siderite in the quartzite beds.

The Lily prospect contains a small lead deposit in a syenitic phase of the monzonite mass on Granite Gulch southeast of Murray. The developments include a few surface cuts and a tunnel 300 feet long on a fault contact between monzonite and Prichard slate. The ore is all on the syenite side of the fault and occurs as stringers and bunches scattered through the outer 20 feet of the igneous rock.

The Jewell prospect, on Cement Gulch north of Prichard Creek, consists of seams and bunches of galena in monzonite. Little work has been done on it, but the occurrence is of unusual interest because it furnishes another specific example of the mineralization following the granitic invasion.

Farther east, in the drainage basin of Paragon Gulch, there are several mines—the Chicago-London, Murray Hill, and Paragon—which although producing some lead are more valuable for zinc and are described in the section dealing with that metal. North of these the Bear Top mine, now idle, is in Prichard rocks near the base of the Burke formation. The ore consists of galena and sphalerite associated with pyrite and chalcopyrite in a quartz-calcite gangue.

On Granite Creek the Giant ledge and the Cedar Creek prospect, both in the Prichard formation, are now being exploited. In previous years each produced some lead ore.

AREA EAST OF POTTSVILLE.

The eastward extension of the area characterized by lead-silver deposits is represented by several prospects and small mines in Montana. The lodes are distributed along the trench determined by the Osburn fault, and in them galena is the most abundant ore mineral, although tetrahedrite is relatively abundant. The usual gangue minerals are siderite and quartz, and in one lode barite is prominent. The Last Chance mine, 3 miles north of Saltese, Mont., with a reported production of $200,000, is the only one that has shipped much ore. The deposits as described by Calkins and Jones[56] include also the Silver Cable, Ben Hur, Bell, Tarbox, Meadow Mountain, Bryan, U. S. Hemlock, and Syndicate.

SLATE CREEK.

South of Mullan several prospects, described by Pardee,[57] have been opened on narrow lodes in the Slate Creek area, tributary to St. Joe River. The lodes are inclosed in the Newland ("Wallace") formation and consist of galena, pyrite, sphalerite, and chalcopyrite sparsely scattered through a siderite gangue. The principal prospects are known as the Silver Spray, Sailor Boy, and Mastodon.

[56] Calkins, F. C., and Jones, E. L., jr., Economic geology of the region around Mullan, Idaho, and Saltese, Mont.; U. S. Geol. Survey Bull. 540, pp. 192–198, 1914.
[57] Pardee, J. T., Geology and mineralization of the upper St. Joe River basin, Idaho: U. S. Geol. Survey Bull. 470, pp. 60–61, 1911.

PLACER CREEK.

Several prospects are situated along Placer Creek about 2 miles south of Wallace. The Vienna-International explores through two tunnels and a shaft a siderite quartz vein from 3 to 5 feet wide inclosed in the middle of the Newland formation. Galena, pyrite, and chalcopyrite occur in scattered bunches. The Castle Rock workings reveal two immense siderite veins inclosed in the Revett quartzite. They strike north of west and dip south. The north vein is about 40 feet wide and the south one about 16 feet. Siderite and pyrite are the dominant minerals, but chalcopyrite and galena occur in scattered bunches and erratically distributed seams. The main Castle Rock vein is probably exposed in the Smart Aleck property a short distance to the west. Here its maximum width is 10 feet. About 1,000 feet west of the Smart Aleck is the Horn Silver property, which contains a quartz-siderite vein about 10 feet wide in which the sulphides, principally pyrite and galena, are very sparsely distributed. This group of deposits is described by Calkins and Jones.[58]

WALLACE AND WARDNER.

The area between Wallace and Wardner contains several deposits which have been worked intermittently since the discovery of mineral deposits in Shoshone County, but none have been continuously profitable. In 1904 none of the deposits were being worked, but in 1916 the Evolution, Yankee Boy, Polaris, Big Creek, and Silverado were worked on a small scale. Most of the lodes are inclosed in the Newland ("Wallace") formation and are characterized by a bunchy distribution of ore high in silver. They strike north of west and dip south.

The Big Creek mine, which has produced about $12,000, is west of Big Creek on a lode that follows the Alhambra fault. The developments comprise about 4,500 feet of work, of which 3,100 feet is in a lower tunnel that enters from the canyon side at an elevation of about 4,000 feet. The ore consists of tetrahedrite, galena, and some argentite and chalcocite in a gangue of siderite and quartz. Locally the ore is almost pure tetrahedrite in veins from the thickness of a knife blade to 2 feet wide. Such material assays from 30 to 32 per cent of copper and from 1,200 to 1,400 ounces of silver to the ton. As shipped the ore runs about 260 ounces of silver to the ton and 5 per cent of copper. Unlike most of the deposits of the group the ore is in the Revett formation.

The Yankee Boy mine, on the west side of the ridge between Big Creek and the head of Polaris Gulch, has been worked intermittently

[58] U. S. Geol. Survey Bull. 540, pp. 199–200, 1914.

for a number of years, and from it about 20 carloads of ore have been shipped. One car averaged 285 ounces of silver to the ton and 9 per cent of lead; others were much lower in silver. The developments comprise a lower tunnel 1,700 feet long and another 500 feet higher, 800 feet long. The ore occurs in two parallel fissures, one of which contains an ore shoot 30 feet long and the other two shoots 80 and 200 feet long. The short shoot, struck in a raise at a point 340 feet above the lower tunnel, is now being worked. It is about 2½ feet in maximum width but feathers out at the ends. The ore minerals are pyrite, galena, and tetrahedrite in a quartz-siderite gangue.

The Polaris mine, east of the Yankee Boy, was being reopened in 1916 after being idle for 16 years. An old shaft is flooded, and only a new development tunnel, 1,500 feet long, is accessible. According to Ransome [59] there are two ore shoots, one 50 feet long and 1 foot in width and another 150 feet long and 6 feet in maximum width. The ore consists of tetrahedrite, pyrite, and galena in a siderite-quartz gangue. Ore as shipped contained about 180 ounces in silver to the ton, 10 per cent of copper, and 8 per cent of lead.

East of the Polaris is the Silverado, developed by a tunnel 4,200 feet long from a point near the floor of the main valley west of Osburn. The vein as seen in a raise near the face of the tunnel is about 10 inches wide, but it is said to be wider in a shaft that was being pumped out at the time of visit. The ore consists of tetrahedrite and a little galena and pyrite, together with quartz and siderite.

Across the valleys of the South Fork of Coeur d'Alene River, about half a mile west of the Silverado, is the Evolution mine, which contains a vein of particular interest because it probably occurs lower in the Prichard formation than any other deposit in the county. Two parallel veins about 150 feet apart strike N. 40° W. and dip 40°–50° SW. They are developed by a vertical shaft, now flooded, 265 feet deep and by a tunnel several hundred feet east of the shaft. The tunnel includes 200 feet of crosscut and 500 feet of drift along the northeast or main vein. This vein is an irregular filling of calcite with subordinate quartz and ranges from 10 to 30 feet in width, as indicated by the openings. Much of the vein is oxidized, but in the last 100 feet of drift primary ore is exposed. It consists of calcite and some quartz through which patches, seams, and bunches of sulphides are sparsely distributed. These consist of pyrite, sphalerite, galena, and chalcopyrite, the last visible only microscopically, in such intimate association that essentially contemporaneous deposition of them can scarcely be doubted from material now available. Pyrite is more widely distributed, however, than the other sulphides, and average material containing it is said to run about $2 in gold and 1.2 ounces

<hr />

[59] U. S. Geol. Survey Prof. Paper 62, p. 188, 1908.

of silver to the ton. The magnetic iron sulphide, pyrrhotite, was not noted in the tunnel but is fairly abundant in certain parts of the dump from the shaft.

ZINC DEPOSITS.

GENERAL FEATURES.

The known zinc deposits of Shoshone County are practically confined to the Coeur d'Alene district and its neighbor, the Pine Creek district. Even in these districts, however, the presence of zinc in the lead ores was long regarded as undesirable and resulted in penalties being imposed on the treatment of the ores. Prior to 1905 no zinc ore was produced in the Coeur d'Alene district. In that year the Success mill of 150 tons began concentrating sphalerite ore which carried 20 to 25 per cent of zinc. In 1906 the Sixteen to One mine (Rex) produced considerable zinc ore, and a little was sorted from the Hercules and shipped. At this time preparations were made at the Helena-Frisco mine to concentrate the zinc ores. The total zinc output of the district in 1906 was 2,054,990 pounds.

The Morning mine became a zinc producer in 1910, and it was here that flotation methods on a commercial scale were first employed in Shoshone County in the separation of galena and sphalerite. Many other lead-zinc mines followed in adding flotation units to their mill equipment, and the zinc production steadily increased and was greatly accelerated in 1914 by high prices for both lead and zinc.

The greatest increase in the zinc production of the district, however, was due to the exploitation of the extensive zinc deposits of the Interstate-Callahan mine, begun in 1913. In that year ore from this mine to the value of $264,644 was extracted, but in 1915 the output was valued at $4,540,671. Since 1915 the production has diminished year by year, the value for 1920 being $1,876,974. The production of zinc in Shoshone County for 1916 was 86,238,283 pounds, valued at $11,555,930; for 1917, it was 77,724,221 pounds, valued at $7,927,871; for 1918, 43,661,314 pounds, valued at $3,973,180; for 1919, 15,994,229 pounds, valued at $1,167,579; and for 1920, 27,932,326 pounds, valued at $2,262,518.

Exploitation of deposits in the Pine Creek district prior to 1916 was greatly retarded because of the difficulty of separating the minerals of its complex zinc-lead ores, and poor transportation facilities. Advances in flotation practice have largely solved the concentration problem. In 1917 railroad surveys were made from Pine Creek Station (on the Coeur d'Alene branch of the Oregon-Washington Railroad & Navigation Co's. line) up Pine Creek to the Highland-Surprise and Constitution mines, and construction began in the fall of that year, but after several miles of grade had been built the work was suspended in 1918. Mining suffered in consequence of this delay;

the Anaconda Copper Co. relinquished its lease on the Douglas mine which it had operated in 1917, and the Highland-Surprise, a large producer in former years, was idle.

The outlook for continued heavy production of zinc ore in the Coeur d'Alene and Pine Creek districts is assured, although in some of the mines, notably the Helena-Frisco, Greenhill-Cleveland, and Success, the bottoms of the ore deposits are apparently close at hand. Production from the Pine Creek district, however, will probably more than offset these losses, and no doubt other deposits will be found. Large areas of the Prichard formation have been but little explored, and the fact that ore deposits have been opened in depth that have no outcrop should encourage exploration in this formation, especially in the country adjacent to known deposits in zones of shearing and faulting.

The zinc deposits of the Coeur d'Alene region occur principally in two areas situated north and south of the Osburn fault. The center of the southern area is about 12 miles west of the center of the northern area along the fault. The northern area extends northeastward from the vicinity of Wallace in the Canyon Creek and Ninemile Creek basins, on either side of exposures of quartz monzonite for 12 miles, nearly to the Bitterroot divide. In this zone are the Helena-Frisco, Greenhill-Cleveland, Success, Rex, Interstate-Callahan, Ray-Jefferson, and Paragon zinc-lead producers, and among the mines whose chief product is lead are the Hercules, Tamarack and Custer, Monarch, and Bear Top mines. The Morning mine, a large producer of both lead and zinc ores, is somewhat isolated from the northern area. The southern area is in the Pine Creek basin and contains the Highland-Surprise, Constitution, Douglas, Little Pittsburg, Nabob, and Northern Light zinc-lead mines and numerous prospects.

Outside these areas, particularly in parts of the Prichard formation, there are several prospects on zinc deposits, but none of them have opened commercial deposits.

These two areas of mineralization, the Canyon Creek and Ninemile Creek basins on the north and the Wardner and Pine Creek districts on the south, are believed to have been adjacent originally, but through cumulative horizontal movements along the Osburn fault the country on the south side was shifted west a distance of about 12 miles. This view [60] was first advanced by O. H. Hershey, geologist for the Bunker Hill & Sullivan Mining & Milling Co., and is supported by the more recent investigation by the writers, as discussed elsewhere (pp. 11–13). An inspection of the geologic

[60] Hershey, O. H., Origin and distribution of ore in the Coeur d'Alene (private publication), pp. 4–5, 1916.

map shows that in places on opposite sides of the Osburn fault synclines are in line with anticlines or vice versa, and that the apparent vertical displacement may range from 1,000 feet to more than 10,000 feet within a short distance. If like structural features in the same formations are brought together, however, the horizontal shift on the south side of the fault is made apparent. This movement was, in part at least, subsequent to the intrusion of the monzonite, as shown by crushed lamprophyre along its plane, and as ore deposition closely followed the intrusion, the faulting in part should be considered subsequent to or concurrent with the ore deposition.

The zinc deposits are contained in Algonkian rocks of the Belt series. In the Coeur d'Alene district, beginning at the lowest formation, the Belt series consists of the Prichard, Burke, Revett, St. Regis, Newland, and Striped Peak formations. In the northern area the zinc deposits are contained principally in Prichard slates and the Burke quartzite, although the Morning lode rocks traverse Revett, but in the southern area the deposits are contained solely in the Prichard formation.

The zinc deposits of the northern area have a significant relation to the areas of monzonite, the outcrops of which extend in a northeasterly direction in the Canyon and Ninemile Creek basins to the Prichard Creek basin. These deposits consist of metasomatic fissure veins and one contact-metamorphic deposit. On the west flank of the intrusive rock the Success deposit, clearly of contact-metamorphic origin, occurs in a tongue of Prichard rocks that extends into the monzonite. North of the Success the Rex and Interstate-Callahan veins lead westward from the vicinity of the monzonite contact. On the east flank of the intrusive mass are the Helena-Frisco and Greenhill-Cleveland lodes. The Helena-Frisco ends at the monzonite, where it partakes of the nature of a contact deposit, the metamorphosed wall rocks containing abundant contact silicate minerals and magnetite being a constituent of the ore.

The Morning lode is several miles east of the nearest surface exposure of monzonite, but a monzonite dike was cut in the adjoining property to the west. This lode well illustrates change in character of ore with depth. Prior to 1910 only the lead-silver content of the ore was saved, but with increasing depth the zinc content increased, and since 1910 it has added materially to the mine revenue. On the lower levels also zinc is more abundant in the west end of the mine— the end toward the center of igneous activity and also of ore deposition—than in the east end.

All the lodes of the northern zinc area have a prevailing westnorthwest direction and generally steep southerly dips. They occupy fissures of shear zones which are not themselves faults of considerable displacement but which were probably formed contemporaneously with the great west-northwest faults of the region.

INTERSTATE-CALLAHAN MINE,

The Interstate-Callahan group lies across a high ridge between the East Fork of Ninemile Creek and Carbon Creek and Missoula Gulch, tributaries of Beaver Creek. The present working tunnel, at an elevation of 4,509 feet, is driven from the Ninemile Creek side and cuts the productive veins at a depth of 1,200 feet below their outcrop. The property was acquired by the operating company in June, 1912, from the Callahan brothers, who had devoted 20 years to the exploitation of the Callahan and Manhattan veins, mainly through tunnels driven from the Carbon Creek side. It is reported that during this time 42 tons of hand-picked ore was shipped whose average content was 70 per cent of lead and 40 ounces of silver to the ton. In order to reach greater depth on the Callahan vein as well as to provide an accessible outlet for the ore the company immediately after acquiring the property began the No. 4 tunnel from Ninemile Creek. In driving this tunnel the Interstate vein was found, and the immense shoot of high-grade zinc ore in this vein had up to 1916 provided most of the revenue of the company. The production increased rapidly after the mine was placed on a working basis. In 1913 the gross value of the ore was $264,644; in 1914, $828,961; in 1915 the Interstate-Callahan surpassed any other mine in the Coeur d'Alene district, with a gross value of $4,540,671, and a net profit of $2,921,488. The production for 1916 continued to be large though not so great as for 1915, the gross value of the ore, largely zinc, being $3,983,522, with dividends paid during the year of $2,789,940. Production decreased gradually during 1917, 1918, and 1920, the gross value of the ore for 1920 being $1,876,974. The decrease in earnings of the company was due largely to the enormous decline in the price of spelter and the abnormally high cost of operating expenses.

Four veins have been cut by the mine developments. All have a general west-northwesterly course and a nearly vertical dip. These veins, which occur in a belt 800 feet wide, are from south to north the Interstate, Callahan, Blue Grouse, and Manhattan. The veins are metasomatic fissure fillings that cut beds of siliceous blue slate of the Prichard formation a short distance west of a large intrusive mass of monzonite. The contact between the slate and monzonite, as shown in No. 4 tunnel, is a fault contact whose general course is north. The slates are greatly contorted near the veins, but their general dip in the tunnel section is at moderate angles to the east. Notwithstanding their proximity to the monzonite mass to the east the slates show little evidence of metamorphism. In view of the general metamorphism of the sediments where they are in intrusive contact with the monzonite this lack of metamorphism in the Interstate-Callahan mine may indicate that the slates were outside the zone of metamorphism at the time of the monzonite intrusion and were subsequently faulted to their present position.

In the east end of the mine, on levels 4, 5, and 6, a monzonite porphyry dike cuts the slate beds, and here an interesting age relation is shown between igneous intrusion, faulting, and ore deposition. This relation is best shown on the 600-foot level, where the dike, from 4 to 15 feet thick, trends nearly east and dips steeply north, making a small angle of intersection with the vein. Both dike and vein are faulted, and the wedge ends of dike and ore are offset along the plane of the fault for 80 feet. Fragments of porphyry and drag ore appear along the fault plane. The dike itself has clearly been replaced by galena and sphalerite and is therefore older than the ore. Near the fault the ore is greatly sheared. The sequence of events was (1) the intrusion of the monzonite porphyry dike, (2) the formation of a fissure across the dike and deposition of the ore, (3) faulting along the dike with a horizontal displacement along the fault plane of 80 feet.

In the west end of the Interstate ore shoot on No. 3 Callahan tunnel level, 800 feet below the surface, a lamprophyre dike about 2 feet wide cuts across the vein, thus proving that it was intruded after the ore deposition.

Of the four veins only the Interstate has been sufficiently exploited to prove the extent of the ore, and curiously enough this vein has no prominent outcrop and its large ore shoot was discovered in driving No. 4 tunnel for the Callahan vein. It was then explored on other levels, the lowest of which, No. 6, proved to have a depth of more than 1,600 feet. The Interstate vein shows a remarkable extension in stope length in depth. On No. 3 Callahan tunnel level it is 700 feet long; on the 350-foot level, 230 feet below No. 3 tunnel, 1,000 feet long; on No. 4 level, 460 feet below No. 3, 1,200 feet long; and on No. 6 level, 450 feet below No. 4, 1,000 feet long. The ore ranges in thickness from 4 to 40 feet and probably averages 12 feet. The Callahan vein, which has a short, pointed ore crest near the surface, expanded to a length of 145 feet on No. 3 level. On No. 4 level the Callahan vein had been explored in June, 1916, for 200 feet only, but it shows strongly at each end of the drift and is from 4 to 10 feet wide. A shaft was being sunk on the vein from this level. On No. 6 level it is only slightly developed. The Blue Grouse and Manhattan veins at the time of visit had only been partly developed on No. 6 level, but it is reported that some ore had been shipped from shallow workings on the Manhattan vein.

The Interstate vein in general is well defined, but veinlets lead off for a short distance from the main ore body into the walls. There is little gouge in the vein and no evidence, except for the fault along the monzonite dike, previously described, of movement subsequent to the deposition of the ores; consequently the walls stand well, and the extraction of the ore does not require the immense consumption of timbers necessary in most of the other Coeur d'Alene mines. The "waste fill" system is used in mining. A floor of 2-inch plank is laid on top of the waste, and the ore as it is shot down is hand sorted and wheeled to the chutes. The waste is obtained from the vein and from waste raises driven into the walls.

The ore of the Interstate vein consists dominantly of clean brown sphalerite of medium grain which breaks in slabs very similar to the blocky slates. Both ore and waste become thickly coated with dust in the stopes, and it is sometimes necessary to wash off this dust in order to separate the high-grade shipping ore from the mill feed and waste. The shipping ore, which is sorted by hand, makes up 10 to 15 per cent of the mine output. It assays about 50 per cent of zinc, and the mill feed, after hand sorting, shows 25 per cent of zinc, 6 per cent of lead, and 2 ounces of silver to the ton. Galena is much more abundant toward the ends of the ore shoot and in veinlets leading into the wall rocks than in the main ore shoot. Pyrite occurs in places in the vein and in the wall rocks but is nowhere abundant. Pyrrhotite was not observed in the ore below No. 4 tunnel, but it was noted in some specimens obtained from the Interstate vein above No. 3 level. The galena occurs as veinlets in sphalerite and was therefore deposited in the vein after the sphalerite. A remarkable feature of the Interstate vein is the small amount of gangue minerals; of these quartz predominates, and it is most abundant at the ends of the ore shoots. A little siderite is commonly present, and in places small seams of calcite and a green mineral, probably chlorite, occur with the quartz. The quartz is of an earlier generation than the ore, for it is replaced by sphalerite and galena.

The Callahan vein contains a much higher proportion of lead to zinc than the Interstate vein, but at the time of visit no stoping was being done on the lower levels and the tenor of the ore was not learned. From the upper workings many carloads of ore were shipped whose average content was 70 per cent of lead and 40 ounces of silver to the ton. The Manhattan and Blue Grouse veins on No. 6 level are composed principally of clean brown sphalerite, but some lead ore was shipped from the upper part of the Manhattan vein..

SUCCESS MINE.

The Success mine, formerly the Granite, is on the southeast side of the East Fork of Ninemile Creek in the long, narrow tongue of

Prichard rocks that extends into the intrusive body north of Burke
known as the Gem monzonite. The Granite claim was located in 1885.
For many years the chief products of the mine were lead and silver,
but since 1906 the value of the zinc output has greatly exceeded that
of lead and silver. The proportion of zinc to lead concentrates now
produced is probably greater than 10 to 1. From 1907 to 1915 the
mine produced ore to an approximate gross value of $2,750,000, and
of this amount $1,250,000 was produced in 1915.

The Success mine [61] is of unusual interest because of the close relation of its ore bodies to igneous intrusion, and it affords the only example of a large contact-metamorphic deposit in the Coeur d'Alene
district. The deposits are exploited by three tunnels and a shaft.
The lowest tunnel and main haulage way, tunnel No. 3, is 700 feet
below the outcrop and extends from the east bank of Ninemile Creek
for 1,200 feet to the shaft. The shaft is sunk about 600 feet below
tunnel No. 3, but the lowest stope is 500 feet below the collar on the
1,200-foot level.

The ore deposits are largely contained in highly metamorphosed
rocks of the Prichard formation, and in much smaller part in monzonite or syenite.

In the vicinity of the mine the sedimentary beds are metamorphosed to biotite schist, micaceous quartzite, and dense greenish
quartzite, and each type contains variable amounts of garnet, pyroxene, biotite, muscovite, and rarely chlorite and epidote. The beds in
general strike northwest and dip 30–85° SW., though in places they
are greatly disturbed and the bedding planes are obliterated. The
sediments are intricately traversed on several levels of the mine by
tongues and dikes of monzonite. In places the monzonite apophyses
are traversed by the ore, examples of which were noted on the 100
and 300 foot levels. Most of the ore, however, is inclosed in mica
schist, a notable amount in the micaceous quartzite, and smaller
amounts in the dense greenish quartzite.

The distribution of the ore bodies is controlled by a zone of shearing
in the biotite schist member of the Prichard, though the ore bodies
are by no means confined to it. The schist member, from 10 to 20
feet thick, strikes a little west of north in the southeastern part of
the mine, but to the north it turns abruptly westward and terminates
against the monzonite in the vicinity of the shaft. Here it dips about
80° S., but toward the southeast it gradually flattens to a dip as low
as 45°. The ore occurs in lenses in the biotite schist and in the micaceous quartzite above and below it. The lenses attain their maximum development near the bend in the formation, and here an outer
and inner series occur—the former principally in the schist and the

[61] Umpleby, J. B., Genesis of the Success zinc-lead deposit, Coeur d'Alene district,
Idaho: Econ. Geology, vol. 12, pp. 138–153, 1917.

latter in the overlying micaceous quartzite. On the west the ore abuts against the monzonite or extends into it a short distance in places, but on the southeast the mineralization dies out through a series of isolated lenses in the same general zone. Most of the lenses in the productive part of the zone are connected either by merging in places along their periphery or by oblique or transverse bodies. Thus if a stope sheet were made by projecting the worked parts of the deposits to a common plane, the ore body would appear as a vein. On the other hand, cross sections and plans of separate levels bring out the essential characteristics of the irregular replacement phenomena. The principal ore body was found in the curved part of the zone and was worked continuously from the surface to the 1,200-foot level. It was largest on the 500-foot level, where the sill floor measured 60 by 100 feet; above and below this level it ranged from 125 to 235 feet in length and from 16 to 18 feet in width.

The ore from the mine, extracted by the shrinkage system without sorting, contains from 8 to 10 per cent of zinc, 2 per cent of lead, and about 1½ ounces of silver to the ton. Sphalerite is the dominant mineral, but galena, magnetite, pyrite, and quartz are present in most places. The characteristic gangue is unreplaced quartzite or biotite schist. The sulphides are distributed through an ore body in blebs, bunches, lenses, and connected areas of ramifying habit. The galena was clearly formed later than the sphalerite, which it traverses in veinlets and connected patches. Pyrite is scattered through the sphalerite in isolated crystals or small groups of crystals and may be contemporaneous with it. Mineralization by pyrite in the surrounding rock, however, is much more extensive than by sphalerite, for in places the hornblende of the monzonite is replaced by pyrite at least 200 feet from any known ore body. Magnetite is intimately associated with the sphalerite and may be nearly contemporaneous with it, although in one of the specimens studied it is distinctly older. Sericite is associated with the sulphides irrespective of whether they replace quartzite, schist, or monzonite. Silicate minerals occur in much of the ore in microcrystalline individuals and aggregates closely intergrown with the sulphide minerals. In the monzonite garnet, pyroxene, a blue amphibole, epidote, chlorite, and rarely tourmaline occur in minute microscopic grains. Garnet is the prevalent contact mineral in the quartzite and schist, but epidote, diopside, and chlorite are present in most places in variable amounts. The essential contemporaneity of the sulphides and silicates is proved by specimens from several parts of the deposit and abundantly bears out Ransome's conclusion [62] that "the association of the ore minerals with the metamorphic silicates is so close that the conclusion of contemporaneous genesis is unquestionable."

---

[62] Ransome, F. L., and Calkins, F. C., The geology and ore deposits of the Coeur d'Alene district, Idaho: U. S. Geol. Survey Prof. Paper 62, p. 185, 1908.

That metamorphism and metallization were subsequent not only to the injection of the magma but to the chilling and solidification of its outer part is proved by the evidence of hand specimens, thin sections, and polished surfaces. One specimen of metamorphosed monzonite cut by an aplite dikelet shows replacement by sphalerite, pyrite, galena, and magnetite and by garnet, amphibole, hedenbergite, and quartz, and both ore and silicate minerals are essentially contemporaneous. The aplite dikelet is itself cut by an intergrowth of pyrite, magnetite, sphalerite, and galena, and in places the feldspar of the aplite is replaced by amphibole, pyroxene, garnet, and sericite.

REX MINE.

The Rex mine (fig. 8) is about 1 mile north of the Success mine, in Prichard slate, on the west side of the Gem monzonite. It has been intermittently productive since 1900, and the value of its output, chiefly in lead and silver, is reported at $700,000. In 1906 it shipped some zinc ore, and after a long period of inactivity it produced in 1916 about 2,200 tons of a zinc-lead concentrate.

Four veins, the North, Rex, Okanogan, and Delaware, crop out on the Rex group and adjoining Patuxent claim, and all but the Delaware, which has not been explored, have been cut by the lower Rex tunnel. These veins are respectively 365, 540, and 470 feet apart on the surface. All have general west-northwest courses, varying a few degrees, and they dip north at angles ranging from 36° to 80°; the North vein has the steepest dip and the Delaware vein the flattest. The Rex vein has produced the greater part of the ore. The mine is opened by two tunnels and an inclined shaft 600 feet deep driven in the footwall of the vein from the lower tunnel. The lowest workings are 1,100 feet below the outcrop of the Rex vein. The Rex vein is a persistent though irregular fissure that cuts blue slate with some intercalated layers of quartzite. In the east drift on the lower tunnel level the vein terminates against the monzonite, and here the slate is altered to hornfels, with the development of metamorphic silicates, such as garnet, pyroxene, and epidote.

The deposit is the result of metasomatic replacement of the slate for the most part, but in places it is a fissure filling characterized by abundant quartz. The ore has been mined principally above the lower tunnel level from two shoots of zinc-lead ore that are probably not over 100 feet long each, but for 350 feet on the lower tunnel level there is a body of zinc ore from 1 to 3 feet wide that will probably be mined if the milling problems are overcome. All the ore is fine grained, and much of it is an intermixture of sphalerite, galena, pyrite, and pyrrhotite. Quartz is the dominant gangue mineral, and siderite occurs very sparsely.

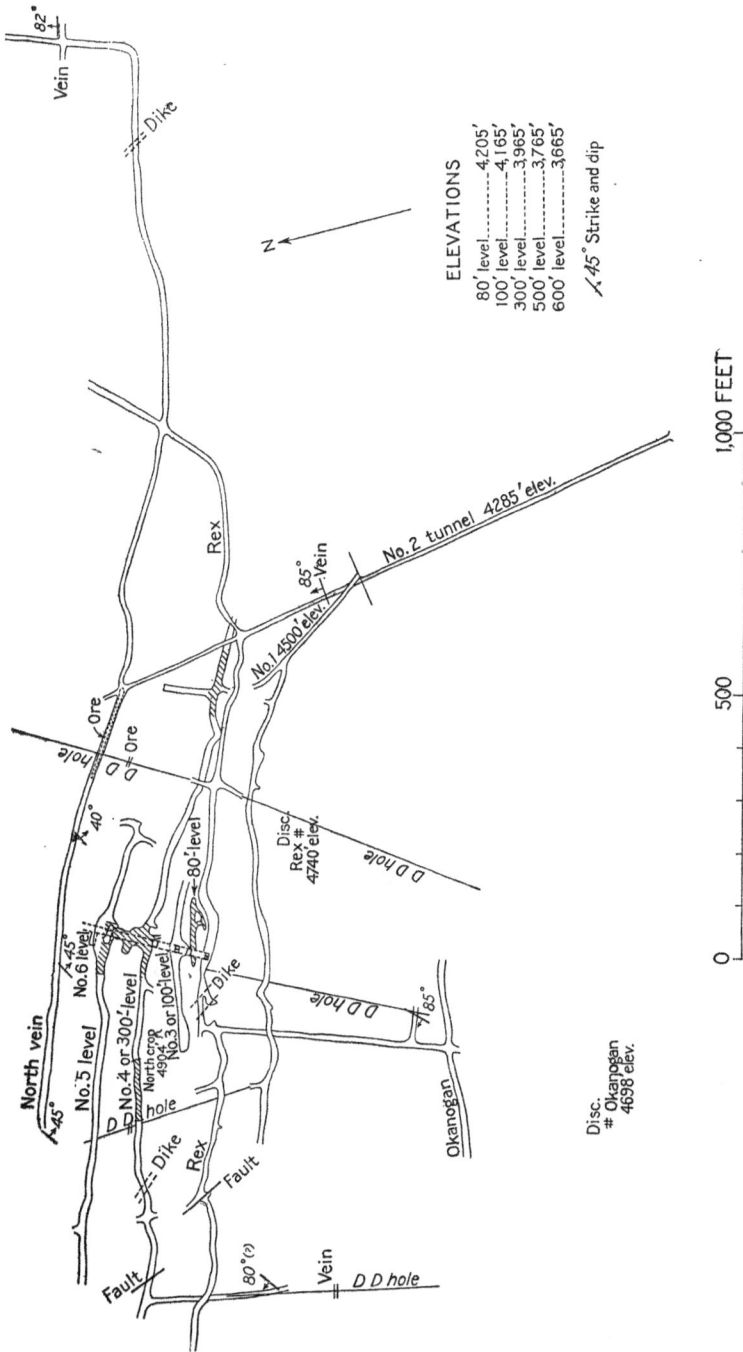

FIGURE 8.—Plan of the Rex mine,

The Helena-Frisco mine is on the south side of Canyon Creek near Gem. It is a consolidation of three mines, the Black Bear, Frisco, and Gem of the Mountains, located on lodes which are apparently separated from one another by faulting and which prior to 1901 were worked independently. The deepest and most extensive workings are those of the Frisco mine, from whose shaft drifts on some of the lower levels explored the adjacent Gem and Black Bear veins. For many years after the location of the claims in 1884 large amounts of high-grade ore were produced, but the records of production were not compiled before 1897. From 1897 to 1906 the mine yielded over 1,000,000 tons of ore from which 100,000,000 pounds of lead and 4,000,000 ounces of silver were obtained. The mine lay idle for many years after 1907, partly because in the lower levels the ore contained so much zinc that it could no longer be treated profitably. The property was acquired by the Federal Mining & Milling Co., and in 1913 work was resumed in the Frisco shaft and a mill adapted to the treatment of the lead-zinc ores was built. From May 1, 1915, to May 31, 1916, 87,363 tons of ore produced 5,149,824 pounds of zinc, 2,639,858 pounds of lead, and 41,792 ounces of silver. The deepest mining is that on the 2,200-foot level of the Frisco shaft. No work has been done in the Gem or Black Bear mines for many years except by lessees, and the workings are for the most part inaccessible.

The Black Bear, Frisco, and Gem veins are contained in quartzite of the Burke formation near an intrusive mass of monzonite which limits the ore on the west in the Gem mine. Whether these veins are independent fissures or are the faulted parts of one vein is a disputed subject. The Black Bear vein lies north and east of the Frisco vein, from which it is separated for several hundred feet by a fault that trends north-northeast, and the Frisco vein is likewise separated from the Gem vein by a fault of similar trend, though the distance between the veins is somewhat less. Near the igneous rock the quartzite is altered to hornstone, but this metamorphism extended irregularly out from the contact, as shown by hornstones in the eastern part of the mine at some distance from known masses of the intrusive rock. Much of the hornstone is black and vitreous; other beds are light green or gray dappled with small aggregates of dark minerals. The microscope shows these rocks to be composed largely of quartz, pyroxene, biotite, garnet, and sericite. The monzonite near the contact contains abundant finely disseminated pyrite, less sphalerite, and locally seams of bladed actinolite and sphalerite.

The three veins of the Frisco mine differ so markedly in character that Ransome considers it possible that they may be distinct veins rather than parts of a faulted vein. All have a general east-west direction and dip steeply south, though the Frisco vein below the 1,800-foot level dips steeply north. The Black Bear vein is a simple fissure filling from 2 to 3 feet wide, the ore of which consists of quartz and galena with practically no sphalerite or pyrite. The Frisco vein, on the other hand, is clearly formed by replacement of the hard quartzite along a fault plane. The maximum width of the ore shoot was 1,000 feet, and the ore was mined for a width of 10 or 12 feet. The vein consists of stringers of galena and sphalerite inclosing much quartzite. In the lower levels of the mine the ore shoot is much smaller and the ore is of lower grade than in the upper workings. There has been little postmineral movement in the vein, and in consequence the walls stand well. The ore of the Frisco vein is made up dominantly of sphalerite and galena with small amounts of pyrite, pyrrhotite, magnetite, and chalcopyrite. Quartz is the dominant gangue mineral. Siderite was not observed in the ore from the lower workings, although it occurred in the upper levels of the mine. The ore now mined contains more zinc and less lead than that from the upper workings. The statistics of production from 1896 to 1907 indicate that the ore yielded 5 per cent of lead and 4 ounces of silver to the ton, with no recovery of zinc. The ore now mined shows a recoverable content of 3 per cent of zinc, $1\frac{1}{2}$ per cent of lead, and less than half an ounce of silver to the ton. The ore of the Gem vein closely resembles the Frisco ore, galena, sphalerite, pyrite, pyrrhotite, and quartz, forming a lode that clearly replaces sheared and fissured quartzite. Siderite is absent or very subordinate. In the Gem vein two shoots of ore converged downward, with maximum length of 800 feet. In lower levels the ore became of too low grade to work. In the upper workings at the time of the visit lessees were mining some high-grade bodies of galena that had been left in the stopes. The ore shoot of the Gem vein ends at the monzonite, in which the vein breaks up into small irregular fractures barren of ore. On the 1,600-foot level west from the Frisco shaft an ore shoot 60 feet long was recently opened which lies in hornstone at the contact with monzonite. This ore consists of clean brown sphalerite with a little galena and is the highest-grade zinc ore in the mine.

The order of deposition of minerals in the Frisco vein is clearly shown with respect to galena, sphalerite, and some of the quartz, but with regard to the other minerals the relations are not obvious. Specimens from the 1,800-feet level show veinlets of steel galena in which there is considerable magnetite inclosing nodules and stringers of a coarser-grained sphalerite, which clearly shows that galena is later than sphalerite. In places stringers of white glassy quartz in the walls apparently do not cut the ore-bearing seams; in other places stringers of quartz are clearly younger than the ore.

The Black Bear and Frisco veins have been cut by lamprophyre dikes. The relation is exceptionally well shown in the Frisco vein in a small stope 35 feet above the 1,600-foot level west from the Frisco shaft. Here a dark fine-grained dike about 4 feet wide cuts through the vein. The contact is frozen, and veinlets of sphalerite end abruptly against the dike. This is particularly noteworthy in view of the occurrence of the disseminated pyrite and sphalerite in the monzonite at the west end of the same ore body, as it places the ore deposition rather definitely in the period of magmatic activity.

MORNING LODE.

The Morning lode is somewhat isolated from the zinc-bearing lodes of the Canyon Creek and Ninemile Creek basins. It crops out on a high ridge between Mill Creek and Grouse Gulch (Pl. XIV), the elevation on the highest part of the ore body being 5,750 feet. The vein is explored to a vertical depth of 3,400 feet, or about 900 feet below the level of Coeur d'Alene River near the portal of the main entrance tunnel. The vein crops out in the St. Regis formation, but as it cuts beds that dip steeply to the east it soon passes into the underlying Revett quartzite, which persists to the lowest level. Several small igneous dikes occur in the vicinity of the Morning lode; in the east end of the mine on the main tunnel level a small lamprophyre dike is associated with a fault that limits the ore on the east. The continuation of the Morning lode to the west as explored by the Star tunnel is cut by a gray porphyry dike believed to be an offshoot from the Gem monzonite, 3 miles farther west. It appears that this proximity to monzonite offshoots offers the best explanation of the high zinc contents of the lode.

The lode trends west and northwest and has essentially a vertical dip. It is a metasomatic replacement deposit that lies in a zone of intense shearing, although it is not closely related to faults of large displacement. There is, however, a zone of intense faulting just south of the ore zone; the Osburn and White Ledge faults are crossed by the main entrance tunnel.

The Morning lode is one of the most persistent in the entire region, and because of its favorable geologic environment it gives promise of yielding ore far below the limits of present exploration. The maximum productive length of the lode is about 2,000 feet (fig. 4, p. 68), but owing to the pinching and swelling of the vein at different points and to irregularities in the pitch of the ore shoots the stoping area is widely different on different levels, although on none is the stope length less than 1,000 feet. The width of the lode ranges from a few feet to 40 feet but averages about 10 feet. In places the vein splits and incloses large horses of country rock. One such horse is 800 feet long and 60 feet in maximum width.

MORNING MILL AND MINE PORTAL.

Photograph by R. A. Standow.

VIEW OF MULLAN, IDAHO.

Continental Divide in the distance.

The branches of the vein that inclose it are about 10 feet wide, but at their junction the stope width is 40 feet. This horse extends from the 200-foot level nearly to the 1,000-foot level. Another horse on the 1,450-foot level is about 300 feet long.

The ore of the Morning lode is valuable for lead, silver, and zinc, but only since 1910 has the zinc been saved. According to common report the zinc content has steadily increased as the lode has been opened to greater depths, but the lack of detailed records of the ore from the upper levels prevents a precise comparison. It is certain, however, that the ore from the west end of the lode contains much more zinc than that from the east end. Galena, sphalerite, quartz, siderite, barite, and calcite are the principal minerals of the ore, but small amounts of pyrrhotite, magnetite, and chalcopyrite occur. The texture of the ore ranges from medium to fine grained and the sphalerite is generally of coarser grain than the galena. Siderite and quartz are the most abundant gangue minerals, but locally barite occurs in irregular bands that may be 1 foot thick and calcite in small bunches and grains intergrown with siderite. Galena, commonly of the steel variety, occurs as reticulating veinlets in siderite and sphalerite. In places the ore veinlets lead off for short distances into the wall rock.

PARAGON GROUPS.

The property of the Paragon Consolidated Mining Co. comprises three groups—the Jewell on the west, the Chicago-London in the center, and the Paragon on the east—which lie across Paragon Gulch, a tributary of Prichard Creek. Work was begun on the Paragon in 1890, but only the Chicago-London has yielded ore; its output comprises 30 cars of ore containing 40 to 50 per cent of zinc and a few cars of lead ore.

The Chicago-London group has four levels, Nos. 2½, 3, 4, and 5, with a vertical difference between the highest and lowest of 350 feet. Tunnels Nos. 2½ and 3 have short drifts on the vein, but No. 4 has about 1,000 feet and No. 5 has 400 feet of drifts. The vein occurs in a fissure which strikes N. 77° W. and dips 70° S. It cuts beds of blue slate and impure sandstone of the upper part of the Prichard formation, whose general strike is north, with steep easterly dip. The ore occurs in rather small and widely separated shoots, two of which, about 300 feet apart on No. 4 level, have been stoped to the surface, and the easterly one is stoped also to No. 5 level. The west stope is about 35 feet long and the east stope probably averages 90 feet. The ore ranges from 1 to 4 feet in width. West of this stope on No. 5 level there are a few bunches of ore, and some ore in the face of the west drift is thought to be at the end of the shoot that was stoped on No. 4 level. Near the east end of the drifts on tunnels Nos.

4 and 5 a vertical north-south fault with gouge 2 to 6 inches thick cuts across the fissure. No ore has been found east of this fault, but the ore does not extend to it from the west, so its relation to the vein is not known.

The deposit is of the metasomatic fissure type, and the ore minerals are sphalerite, galena, and pyrrhotite in a gangue consisting predominantly of quartz but also containing calcite and siderite. Beyond the ore shoots the vein is composed principally of quartz. Some of the ore is a mixture of steel galena and fine-grained sphalerite in which are fragments of unreplaced wall rock, but probably the greater part is composed dominantly of coarse-grained sphalerite. Galena is much more abundant toward the ends than in the middle of the ore shoots, a relation similar to that observed in the Interstate vein. Quartz was clearly the first mineral deposited in the fissure; it was replaced by sphalerite and galena, in part deposited at the same time but for the most part the lead later than the zinc, as shown by numerous seams of galena cutting sphalerite. The age relations of calcite, siderite, and pyrrhotite are not apparent from the specimens collected.

The Paragon or Chicago-London vein is strikingly similar to the zinc-lead veins of the Pine Creek district. It has a similar strike and dip; it occurs near the same stratigraphic horizon; both are of the metasomatic fissure type; and the kinds and association of vein minerals are identical, though there is, of course, considerable variation as to the relative abundance of any ore mineral, such as pyrrhotite, calcite, or siderite.

A lamprophyre dike rich in biotite occurs in the hanging-wall side, 115 feet from the ore on tunnels Nos. 3 and 4, but converges toward the vein fissure on No. 4 level, 450 feet west of the point where the crosscut tunnel intersects the vein.

The Paragon vein is a narrow zone of fissuring that strikes N. 70° W. and dips 45° S. The principal ore mineral is galena, which occurs in bunches at points where the fissure cuts sandstone beds inclosed in the slate series. No continuous ore shoot has been found. The gangue is quartz and siderite, and the bunches of ore minerals are for the most part metasomatic replacement deposits.

The development work on the Jewell group is done in one of several small areas of monzonite that lie west of the Paragon. Here there are seams of galena, from a knife edge to 1 inch thick, that trend in different directions and at their intersections make small bunches of ore. The occurrence of lead in the monzonite is of unusual interest because it furnishes another specific example of mineralization following the intrusion.

MURRAY HILL MINE.

The Murray Hill mine, formerly the Black Horse, has produced considerable ore from a vein that strikes N. 70° W. and dips 65° S.

It cuts Prichard slates, which strike N. 20° W. and dip steeply northeast. The ore occurs principally in two shoots 400 feet apart, one of which has been stoped for a length of 80 feet and for 175 feet above the tunnel, and the other stoped for 70 feet along the drift and for 170 feet above it. The average width of the shoots is about 2½ feet, but they wedge irregularly to seams at their ends. The central parts of the ore shoots contain the highest-grade ore, composed in large part of sphalerite and a little galena associated with quartz and calcite gangue. About 120 feet north of this vein a parallel vein has been followed for 1,000 feet. A shoot of ore 35 feet long and 2½ feet wide was found in this vein. On the margins of the shoot galena, pyrite, and sphalerite occur, but in the center the sphalerite is free from other ore minerals. Quartz and calcite comprise the gangue, with a very little siderite. The sequence of deposition is in the main quartz, calcite, sphalerite, and galena, with pyrite, followed probably by a second generation of calcite.

TERRIBLE EDITH MINE.

The Terrible Edith mine is near the head of Wesp Gulch, northeast of Murray. It has produced ore valued at $60,000, principally in zinc. The vein strikes N. 32° W. and dips 40° SW. It is opened for 825 feet on the lower tunnel level, where two ore shoots about 100 feet apart have each been stoped for 60 feet on their strike. The ore body is from 1 to 5 feet wide in these stopes. Quartz is the dominant gangue mineral and was the first mineral introduced into the vein, as it is replaced by both sphalerite and galena. The galena also cuts the sphalerite in small seams, which is the prevailing relation in most of the ore deposits in the Coeur d'Alene district.

DEPOSITS OF THE PINE CREEK DISTRICT.

The Pine Creek district, which has been described more fully in a separate paper,[63] comprises an area that lies in the drainage basin of Pine Creek southwest of the Wardner subdivision of the Coeur d'Alene district. Many prospects were located in the Pine Creek district soon after the discovery of ore in the Coeur d'Alene district, but except at one or two little work was done on them until the last few years, when high metal prices stimulated their development and resulted in a considerable production of zinc-lead and antimony ores. The zinc-lead producers comprise the Highland-Surprise, Constitution, Douglas, Hypotheek, Nabob, and some other less important properties. The Highland-Surprise mine produced ore to the value of about $30,000 in 1912. It lay idle for several years but in 1915 began active operations and in 1916 was the largest producer of lead-

---

[63] Jones, E. L., jr., A reconnaissance of the Pine Creek district, Idaho: U. S. Geol. Survey Bull. 710, pp. 1–36, 1919.

zinc ore in the Pine Creek district. It continued to be one of the larger producers in 1917, but in 1918 was idle, mainly because of poor transportation facilities and higher costs of operation. Lead-zinc ore was shipped from the Constitution and Douglas properties in 1916. A 150-ton mill was built at the Constitution mine during that year and put into operation early in 1917. The Douglas mine was leased and operated by the Anaconda Copper Co. in 1916 and began shipping crude ore which was treated in the company's electrolytic smelter at Great Falls, Mont. The company continued to operate the mine in 1917, it furnishing a gross value of $346,718; in 1918, however, the company relinquished its lease on the mine. The Hypotheek mine, one of the old properties of the Pine Creek district, produced in the six months period of June to December, 1916, ore to the value of about $90,000, and production was continued in 1917 and 1918. In 1917 the Nabob property produced some lead-zinc shipping ore and lead concentrate of a good grade, and a few tons of lead concentrate came from the Wyoming property. The mines of the Pine Creek district, which considerably lessened their production in 1918, did not produce in 1919, as the railroad into the district, begun in 1917, has never been completed. The output of the district has always been limited by poor transportation facilities. Railroad surveys were made in 1917 from Pine Creek station (on the Coeur d'Alene branch of the Oregon-Washington Railroad & Navigation Co.'s line) up Pine Creek to the Highland-Surprise and Constitution mines, and construction was begun in the fall of that year, but after several miles of grade had been built the work was suspended and as above stated has never been completed. The Highland-Surprise, Constitution, Douglas, and a few other mines are in a position to ship ore if the conditions of transportation are improved. At the Nabob mine a 150-ton mill was under construction in 1919 for the concentration of ore and some development work was being done.

No accurate estimate of the zinc production of the Pine Creek district can be made now.

The zinc-lead deposits are metasomatic fissure veins that occur in slate in the upper part of the Prichard formation. Most of these deposits occur in zones of intense shearing, and in such deposits the replacement of the country rock by ore is much more complete than in the few deposits where fissures were formed under less intense pressure, so that part of the ore could be deposited as a fissure filling. Some of the veins are closely associated with faults of large displacement: the Highland Surprise vein makes a small angle with the Placer Creek fault, and the Constitution vein is closely parallel to the Pine Creek fault. The zinc deposits in the Pine Creek district are confined principally to a block between the Placer Creek and Pine Creek faults, in which are the Constitution and Douglas mines, and to a zone extending northwestward from the Highland-Surprise to the Northern Light mine. On the West Fork of Pine Creek there are many other zinc prospects in the Prichard formation, but none of them has opened an ore shoot.

The Prichard formation occupies the central part of the Pine Creek basin and in general forms an anticline, though it is modified by faulting. The Burke and Revett rocks flank the Prichard formation on the east and south and partly on the west. To the east these formations are separated by a fault, to the south they overlie the Prichard west of a point near the forks of Trapper Creek, but east of Trapper Creek these formations are brought into contact by the Pine Creek fault. West of the West Fork of Pine Creek the slate beds are in part conformably overlain by quartzite.

No economic deposits have to date been encountered in the younger quartzite which flanks the Prichard on the south and west, and this is probably due to lack of faulting and zones of intense shearing.

The Constitution and Douglas lodes strike northwest and dip southwest. These lodes are clearly of the replacement type. They occur in greatly sheared slate, and the texture of the ore is very fine grained, like that of the parent rock. The ore shoot of the Douglas mine as exposed on the lowest tunnel level, which attains a depth of 300 feet, is over 800 feet long, and has an average width of 4 feet. The Constitution vein contains two ore shoots on the tunnel level; one is 400 feet long and the other has been explored for 265 feet. The ore ranges from a few inches to 5 feet in width.

The ore of the two mines is very similar and consists of an intimate mixture of fine-grained sphalerite and galena associated with smaller amounts of pyrite and pyrrhotite. Commonly fragments of unreplaced slate occur in the ore. The replacement of slate by sulphides is complete in many parts of the vein, but toward the ends and walls of the ore shoots the sulphides are sparsely disseminated. The deposition of sphalerite and galena was contemporaneous for the most part, as a thin section of lean ore shows these minerals to be intimately intergrown, but in places seams of galena traverse the fine-grained composite ore, showing that the deposition of galena continued later than that of sphalerite. Pyrrhotite, though not everywhere observable to the eye, may be detected in some of the finely crushed ore by passing a magnet over it. Gangue minerals are not abundant in the Constitution and Douglas lodes. Quartz predominates, and it is accompanied by a little calcite. Siderite was not observed in the ore.

The mineral zone that extends northwestward from the Highland-Surprise mine is composed of a number of lodes of moderate length, which strike in general west-northwest and dip south. The Little

Pittsburg lode, however, is an exception to the general trend of veins in this zone, for it strikes N. 15° W. This zone comprises both metasomatic veins and those in which replacement was subordinate to fissure filling. Thus the Highland-Surprise, which lies near the Placer Creek fault, is a replacement deposit, but deposits progressively farther from the fault partake more and more of the aspect of fissure fillings and present differences both in texture and mineral composition. The ore is coarser grained, quartz is much more abundant, and locally there is a little siderite. Galena is generally more abundant, particularly where the wall rocks are quartzite, and chalcopyrite is a common though variable constituent of the ore. Pyrrhotite is abundant in parts of the Highland-Surprise, Little Pittsburg, Nabob, and Northern Light veins and in places predominates over the other vein minerals. Its distribution is somewhat erratic, however, and its scarcity in the Constitution and Douglas lodes and its abundance in the Highland-Surprise and Nabob veins are difficult to account for, as they were apparently formed under similar conditions.

The extent of the ore shoots along this zone is little known, but several show promise of producing large quantities of ore. The Highland-Surprise and Nabob veins have received the greatest amount of development. In the Highland-Surprise the ore occurs in shoots of moderate length, one of which is 140 feet long and has been stoped through a vertical distance of 225 feet. The Nabob vein has been cut at a vertical depth of 400 feet, but no large ore body of good grade has been developed in it. A vein on the adjoining Denver claim, however, shows in a shallow tunnel an ore shoot 140 feet long and from 2 to 4 feet wide, that is said to contain 31 per cent of lead, 20 per cent of zinc, and 19 ounces of silver to the ton. The Little Pittsburg vein is explored by a tunnel for 200 feet. The ore is from 4 to 21 feet wide and in its widest part is reported to assay 8 to 10 per cent of zinc and 5 per cent of lead. The Northern Light mine on the 400-foot level encountered a zinc-lead vein about 9 feet wide which on the surface showed only barren quartz stringers. Drifts on the ore shoot extend for 200 feet, but its length is still undetermined. The mill feed assays 6 to 10 per cent of zinc, 3 to 8 per cent of lead, and 1½ to 3 ounces of silver to the ton.

The Liberal King vein, midway between the Highland-Surprise and Northern Light mines, is probably the best example of a fissure filling in this zone. The vein strikes N. 75° W., is vertical, and cuts beds of blue slate that strike N. 20° W. and dip 70° E. It is explored for 500 feet by a drift on its north wall and by crosscuts to the south wall at intervals of 100 feet. The vein is about 7 feet wide and generally has well-defined walls. Quartz is the dominant vein filling and in places is practically free from sulphides through a width of several feet. In most places, however, it contains pyrite, sphalerite, chalcopyrite, and galena, named in the order of their abundance. The ore contains a little gold, which is probably in chalcopyrite. Siderite was not observed in the vein, and pyrrhotite is absent or only sparingly present. No stoping has been done on this vein.

COPPER DEPOSITS.

The copper-bearing deposits of Shoshone County are widely distributed and occur under varying conditions, but the greater part of their production has been derived from one mine—the Snowstorm, of the Coeur d'Alene district. This deposit, after yielding approximately $10,500,000 in copper, silver, and gold, was abandoned in 1915 after much fruitless search for ore below a fault against which the main mass terminated. Since that time the copper production of the county has come chiefly from siderite veins and as a by-product from the lead-silver veins of the so-called dry belt between Wallace and Wardner and certain mines of the Wardner district, particularly the Caledonia. Production from these sources, though important, will probably never equal the average yearly production of the Snowstorm mine during its period of activity. From 1884 to 1916, inclusive, Shoshone County produced 64,500,000 pounds of copper.

RELATION TO ROCKS OF DIFFERENT CHARACTER.

The copper deposits of Shoshone County occur in Algonkian rocks of the Belt series, in formations ranging from the Prichard, at the base of the Coeur d'Alene section, to the Newland ("Wallace") formation. So far as known copper deposits have not been found in the Striped Peak formation, at the top of the section. Deposits too small to be of commercial importance occur in fracture zones in diabase in the St. Joe River basin.

FORM AND DISTRIBUTION.

The copper-bearing deposits may be classified according to their mode of occurrence as siderite veins, disseminated deposits, lead-silver veins carrying copper, contact deposits, and brecciated zones in diabase.

The siderite veins are widely distributed, but they are most prominently developed in a zone several miles wide which extends from the Bitterroot divide east of Adair in a west-northwesterly course across the county and through the Pine Creek basin, a distance of 30 miles. Another belt of siderite veins occurs along St. Joe River southeast of Avery, and scattered veins occur in the drainage basin of the North Fork of Coeur d'Alene River.

Disseminated copper deposits occur in the vicinity of the Snowstorm mine, and they are reported [64] to extend for 7 miles in beds of

---

[64] Huston, George, The copper beds of the Coeur d'Alene: Min. and Sci. Press, vol. 110, pp. 145–147, 1915.

the formations that form the Granite Peak syncline. Deposits of this character are reported to occur in the St. Regis, Newland, and Revett formations. In only one of these formations, however, the Revett, have deposits of economic importance been developed, principally in the Snowstorm mine. The National mine, which is in the same ore-bearing formation about 1½ miles west of the Snowstorm, contains much leaner ore, and its operations were conducted at a loss in 1916. North of the National mine no deposits of importance have been found.

The Caledonia mine is the principal lead-silver mine that produces copper as a by-product, but the Yankee Boy, Polaris, Big Creek, and Silverado mines, in the mineral belt between Wallace and Wardner, also produce small quantities. The copper is mainly in the form of chalcopyrite and tetrahedrite, and a high silver content is invariably associated with tetrahedrite in this belt. The Hypotheek mine and several prospects in the Pine Creek district and the Black Bear Fraction on Canyon Creek contain copper in notable quantities but hardly enough to warrant saving as a separate product.

A copper deposit on St. Joe River near the mouth of Black Prince Creek occurs at the contact of monzonite and metamorphosed strata of the Newland formation and is the only known example of this type of copper deposit in the county.

In the upper St. Joe River basin brecciated zones in diabase of the Wishards sill contain chalcopyrite in small amounts associated with siderite, but deposits of this type differ only in wall-rock environment from those described as siderite veins.

SIDERITE VEINS.

The siderite veins are widely distributed throughout the county but are most abundant south of the South Fork of Coeur d'Alene River, and in this southern area the veins are divided into two groups for convenience in description. One group occurs in the area south of Coeur d'Alene River to a line drawn westward from the Bitterroot divide near Adair; the other group comprises the veins in the upper St. Joe River basin. In the northern part of the county the Empire Copper Co. is exploiting a large siderite vein, and a short distance west of the county line on Little North Fork several other properties are located on siderite veins.

SOUTHERN AREA.

In the southern area the veins have a dominant west-northwest trend parallel to the great faults of the region, the most prominent of which are the Osburn and Placer Creek faults. From the Bitterroot divide westward to the head of Placer Creek and in the upper

St. Joe River basin the veins are coextensive with the great intrusive mass of diabase known as the Wishards sill.[65] In the upper St. Joe River basin this sill has been repeatedly faulted by faults south of the Placer Creek fault, which cause the sill to crop out in parallel bands, the most southerly of which follows the bluff for several miles on the south side of the river. Many prospects are situated on these veins, but only a few have produced ore. The principal producers are the Richmond and Monitor mines, near the Bitterroot divide. The country rock of most of the veins belongs to the Newland formation, but some are in the St. Regis and Revett, particularly those in the Pine Creek drainage basin. The veins are well-defined nearly vertical fissure fillings in which the gangue is chiefly siderite, together with some other carbonates and a little quartz, and the primary ore is chiefly chalcopyrite and pyrite.

The veins range from a few inches to 20 or 30 feet in width, and although they are for the most part fissure fillings they replace the wall rocks to some extent. The priority of siderite to quartz is clearly shown and is particularly noticeable along the Bitterroot divide, where quartz veinlets cut siderite. In some veins perfect siderite crystals, commonly 1 or 2 inches in diameter, are partly embedded in quartz that has filled cavities lined with druses of the carbonate.

The Monitor mine,[66] near the Montana line, on the Bitterroot divide, has shipped about 500 tons of ore but has been idle since 1910, when the hoist and buildings were destroyed by fire. The vein matter as exposed at the surface is a typical gossan—a soft porous mass consisting mainly of limonite with some quartz and a little malachite. The principal vein is about 15 feet thick and stands nearly vertical. Unoxidized ore shows chalcopyrite and pyrite in a gangue of siderite, calcite and quartz.

The Richmond mine is on a vein a short distance north of that exposed in the Monitor mine. It is developed by several shafts on the Bitterroot divide and by a tunnel driven from the Idaho side which attains a depth of 350 feet. The mine was actively productive in 1916 and during the period between September 13 and October 20 shipped 17 carloads of ore which yielded $20,000 net. One carload is said to have assayed $15\frac{1}{2}$ per cent of copper and $4 to $5 in gold to the ton. The average tenor of the ore is said to be 10 per cent of copper. The vein strikes N. 75° E., dips steeply north, and is from 5 to 10 feet wide. At the time of visit in 1912 oxidation extended to the deepest point to which the vein had been explored, 175 feet below the outcrop. The oxidized part of the vein consists of limonite, derived from siderite, traversed by veins of quartz with some copper carbonate.

[65] Pardee, J. T., U. S. Geol. Survey Bull. 470, p. 39, 1911.
[66] Calkins, F. C., and Jones, E. L., jr., U. S. Geol. Survey Bull. 540, p. 196, 1914.

West of the Richmond the Manhattan prospect, on Manhattan Creek, and the Alice and Alpina prospects, on Kelley Creek, explore veins from 2 to 10 feet thick which strike nearly east and dip 65°–70° N. Pyrite and chalcopyrite are the principal ore minerals and occur in a gangue of siderite, quartz, and calcite.

In the vicinity of Stevens Peak several prospects have been extensively developed but have produced very little ore. The Park prospect, south of Stevens Peak, is on a zone of east-west fissuring in the Newland formation. The ore consists of pyrite and chalcopyrite in a gangue of siderite and quartz, and it is said to carry appreciable quantities of gold and silver in addition to the copper.

North of Stevens Peak on Willow Creek the Reindeer Queen group is extensively explored by two tunnels, of which the lower one attains a depth of 1,200 feet below the outcrop of the vein. In this tunnel the vein is exposed for 1,600 feet in a drift. It strikes east, dips about 70° S., and is inclosed in the St. Regis formation. A large basic dike, which at the west end intersects the vein at an acute angle, is apparently later than the ore, although the vein has not been found west of it. The vein is from 2 to 20 feet wide and is essentially a simple fissure filling, though in places it splits into several parts. There has been considerable movement along the walls of the vein, with the development of chlorite and sericite. The vein matter consists mainly of coarse-grained siderite, calcite, and quartz. Through this gangue chalcopyrite is irregularly distributed in grains, seams, and masses, some of which are a few inches in diameter. Parts of the vein contain ore, but the tenor of the deposit as a whole is low. Two carloads of hand-jigged copper ore are reported to have been shipped in 1914. Assays of ore in the bin are said to have yielded an average content of 3 per cent of copper and 0.6 ounce of silver to the ton.

A short distance north of the Reindeer Queen is the Carney Copper group, located over an east-west shear zone in rocks of the St. Regis formation. The lode is developed principally by two tunnels 425 feet vertically apart. It is from 10 to 15 feet wide and is well defined by walls that dip steeply north. Pyrite, magnetite, chalcopyrite, and galena constitute the primary ore minerals of the vein. Chalcocite is a secondary product that partly replaces chalcopyrite. The gangue minerals are siderite, barite, calcite, and quartz. Barite occurs as veins in places 4 feet wide. It is fine grained and commonly shows banded structure, with the development of small pyrite crystals. The distribution of ore minerals is irregular. In the upper tunnel chalcopyrite occurs in seams and bunches in quantity sufficient to form milling ore for a distance of 200 feet. In the lower tunnel chalcopyrite occurs sparsely, but magnetite is abundant locally and a little galena replaces calcite and barite, though the material is hardly rich enough to be regarded as ore.

Several prospects[67] have been opened along Placer Creek on or near the Placer Creek fault. The Vienna-International group, near the mouth of Flora Gulch, is developed by two tunnels and by a shaft, inaccessible at the time of visit, sunk from the lower tunnel level. An east-west siderite vein from 3 to 5 feet wide and of vertical dip is the basis of the prospecting. West of the shaft the vein is apparently cut off by a northwestward-trending fault. The ore on the dump, which apparently came from the shaft, shows scattered bunches of galena, pyrite, and chalcopyrite in a quartz-siderite gangue, but the material is apparently too poor to be concentrated at a profit.

The Castle Rock prospect, west of the International, is on the south side of Placer Creek, in a narrow tongue of the Revett quartzite that extends into rocks of the Newland formation. Both contacts of the Revett and Newland formations are fault contacts, the Big Creek fault making the north boundary and the Placer Creek fault the south boundary of the Revett quartzite. A tunnel driven 565 feet southwestward penetrates the Revett quartzite and near the end cuts through the Placer Creek fault. Two siderite veins are exposed in the tunnel. One near the entrance strikes N. 65° E. and dips 75° S. It is 95 feet wide as seen in the section but probably about 40 feet wide at right angles to its strike. It is composed mainly of siderite, quartz, and pyrite but contains innumerable inclusions of quartzite which have been partly replaced by the gangue minerals. Scattered bunches of chalcopyrite were noted, but the tenor of the vein as a whole is low in copper. Galena is reported from it. A shipment of 32 tons of selected ore is said to have yielded 2.61 per cent of copper, 3 per cent of lead, 3 ounces of silver, and $2.75 in gold to the ton. The other vein is in the Revett quartzite near the Placer Creek fault. It strikes N. 50° W., is nearly vertical, and is about 16 feet wide. Its mineral composition is similar to that of the northerly vein but in addition it contains a little gray copper in seams.

The Smart Aleck and Horn Silver prospects, near the forks of Placer Creek, exploit veins that are situated similarly in a narrow tongue of the Revett quartzite between the Big Creek and Placer Creek faults. The veins strike N. 60° W. and dip steeply south, and each has a maximum width of 10 feet. Quartz and siderite are the dominant minerals of these veins, with pyrite and small amounts of irregularly disseminated chalcopyrite.

In the southern part of the Pine Creek basin manganiferous siderite veins occur, some of which contain small amounts of disseminated chalcopyrite, but no commercial deposits have yet been

---

[67] U. S. Geol. Survey Bull. 540, pp. 199–200, 1914.

discovered in them. The Colusa, Black Diamond, Palisade and other prospects have been located on these veins. The veins are fissure fillings from a few inches to 40 feet wide. They strike in a general west-northwest direction and cut the Burke, Revett, and St. Regis formations, which in the southern part of the Pine Creek basin are undisturbed by major faults. The zone of fissuring is several miles long, and it probably extends beyond the Pine Creek basin both to the east and the west. The outcrops of these veins are oxidized to iron and manganese oxides and quartz. An assay across the oxidized portion of a vein of the Black Diamond group is said to have yielded 19 per cent of manganese.

DEPOSITS IN DIABASE ALONG ST. JOE RIVER.

Between St. Joe River east of Avery and the Bitterroot divide many siderite veins accompany a zone of west-northwest faulting. Many of these veins cut or are adjacent to the large intrusive mass of diabase known as the Wishards sill,[68] which is repeated several times by means of the northwesterly faults. Other veins cut smaller diabase dikes. In the diabase the fissures are poorly defined but may be re- solved into brecciated zones. Though accompanied by more or less gangue minerals the copper minerals occur both in the fissures, where the inclosing rocks are sedimentary, and in the sheared zones in dia- base; hence the two types of mineralization are not sharply separable except as to wall-rock environment. Gold accompanies the copper in both. The descriptions of prospects in this area are largely taken from the report by Pardee, cited above.

The Ward mine is near the southeast spur of Wards Peak, on the Idaho-Montana boundary, and lies within the two States. The country rock consists of green shale and quartzite of the lower part of the Newland formation, which are cut by a vertical dike of dia- base trending about N. 75° W. A mineralized shear zone 50 feet wide trending west-northwest affects the diabase principally. Within it are numerous seams and veins of quartz, calcite, and siderite, carrying small amounts of chalcopyrite, pyrite, and chalcocite. The zone is said to assay about $4 in gold to the ton, but the average cop- per content is not known.

The St. Joe quartz prospect, on St. Joe River near the mouth of Bluff Creek, is in a mineralized shear zone in the Wishards sill. The zone trends about east and contains irregular seams and bunches of quartz and calcite with a little chalcopyrite, but no ore shoot has been developed.

In the vicinity of Conrad's crossing, on St. Joe River, several pros- pects are located on small veins and shear zones in the Wishards

68 Pardee, J. T., U. S. Geol. Survey Bull. 470, p. 47, 1911.

sill. The mineralization consisted in the formation of minute crystals of chalcopyrite and pyrite along shear planes and the development of seams and irregular bunches as much as 3 feet wide of calcite, quartz, and siderite carrying chalcopyrite and pyrite.

Near the mouth of Bird Creek a large siderite-calcite vein which strikes N. 30° W. and dips 35° SW. is explored by a short adit. No copper minerals were noted in the vein, but it contains considerable pyrite.

VEINS NORTH OF COEUR D'ALENE RIVER.

North of Coeur d'Alene River copper-bearing siderite veins are most abundant in the drainage basin of the Little North Fork, but most of them, notably the Handspike, Hamburg-American, Riverside, Horseshoe, Idaho Queen, and Granite properties, are situated on veins a short distance west of the county line. The principal producer, however, the Empire Copper Co., exploits a vein on Little North Fork just east of the county line. These veins are similar to those in the southern area. They trend in general northwest, and most of them appear to have been formed incidentally to northwest faulting. The Empire Copper vein is inclosed principally in the Revett quartzite; the country rock of the other veins is dominantly of the Newland formation. The veins range from a few inches to 50 feet in thickness. They are composed principally of quartz and siderite. Most of them contain shoots of ore consisting of chalcopyrite and pyrite in small veins and bunches or disseminated. Near the surface the veins are oxidized and consist of crusts of quartz in a spongy mass of iron and manganese oxides. In addition to the copper content of the veins, gold to the value of $4 a ton is reported in some of them.

The vein on the Empire group strikes N. 54° W. and dips 53° SW. It was discovered in 1886 but was little exploited until recently. In 1916 five carloads of ore carrying 20 per cent of copper were shipped and a mill with a capacity of 150 tons a day was built. The vein is about 12 feet wide and consists principally of quartz and siderite, with chalcopyrite and pyrite as the metallic sulphides. The better ore contains about 50 per cent of copper and occurs in a shoot 40 feet long near a premineral fissure which trends northeast. The average content of the lode is said to be about 1½ per cent of copper. The ore is oxidized to a depth of 100 feet, with the development of malachite, azurite, cuprite, bornite, and rare chalcocite. Iron oxides are abundant, but there is apparently little manganese.

DISSEMINATED DEPOSITS.

SNOWSTORM LODE.

The Snowstorm lode, unique as a large producer of copper ore in a region chiefly important for lead, silver, and zinc, was first exploited in 1903 and ceased to be productive in 1915. The lode is opened by four tunnels, the lowest of which, No. 4, at an elevation of 4,750 feet above sea level, attains a depth of more than 1,600 feet below the outcrop.

The Snowstorm copper deposit is a zone of disseminated copper minerals in a stratum of medium-grained hard white quartzite of the Revett formation. This formation is on the eastern flank of the Granite Peak syncline, and it extends from the mine in a northwesterly direction for many miles. The Newland and St. Regis formations, on the eastern flank of this syncline, also contain disseminated copper minerals, but up to date no deposit of commercial importance has been developed in them. The northern limit of the copper-bearing stratum containing the Snowstorm lode is not definitely known but is apparently penetrated by the workings of the Snowshoe, Missoula Copper, and National mines, and if so, the stratum has a length of about 1½ miles. The southern limit of the copper-bearing stratum is the Snowstorm fault,[69] which apparently cuts off the ore body. The Snowstorm fault is more properly a fault zone in which the largest and most northerly fault is that which cuts off the ore and brings different beds of the Revett into contact. Other faults in this zone bring the St. Regis beds into contact with the Revett, the relation being that of a reverse or thrust fault.

The Snowstorm lode is in a hard quartzite stratum which strikes N. 60° W. and dips 65° S. Stoping has been carried on along a maximum length of 700 feet in this stratum and through a width of 40 feet. No ore has been found below the 900-foot level. The ore shoot has a nearly vertical pitch near the surface, but its course quickly changes to a low eastward pitch, and as the strike of the bedding and main fault converge toward the east it happens that the ore shoot gives out at less depth than at a point directly beneath the outcrop.

The principal unoxidized ore of the Snowstorm mine consists of quartzite impregnated with minute particles of bornite, chalcocite, chalcopyrite, and tetrahedrite. The richest ore is of a uniform dark-gray tone; the leaner ore is lighter and finely dappled. Oxidized ores largely predominated in the Snowstorm mine, and these were stoped to the 600-foot level, although in places the oxidized zone extends to the 1,600-foot level. Malachite is the chief product of the oxidation of the sulphides; cuprite occurs in smaller amounts. The oxidized ore was formerly treated by a leaching process, and the sulphide ore was finely crushed and concentrated. The finely disseminated sulphides when studied microscopically are seen to replace the siliceous cementing substance of the quartzite, sericite flakes, siderite,

[69] Ransome, F. L., and Calkins, F. C., U. S. Geol. Survey Prof. Paper 62, p. 151, 1908.
Calkins, F. C., and Jones, E. L., jr., U. S. Geol. Survey Bull. 540, p. 206, 1914.

and in much smaller part the grains of quartz and feldspar. The average tenor of siliceous ores shipped to smelters in 1904 was 4 per cent of copper and 6 ounces of silver and 0.1 ounce of gold to the ton. In 1912 the concentrating ore contained 2.75 per cent of copper. The sulphides were probably introduced through minute fractures in the quartzite and were deposited by upward-circulating waters.

NATIONAL LODE.

The National lode, east of Mullan (Pl. XV), is developed by a tunnel over 4,000 feet long which reaches a vertical depth of 1,200 feet. The tunnel penetrates the Newland and St. Regis formations and extends into the ore-bearing stratum of the Revett quartzite. Several faults were cut. The largest, with east-west course and steep south dip, separates the St. Regis and Revett formations and probably limits the ore, though at considerable depth below the present tunnel level.

The ore-bearing stratum strikes northwest and dips about 45° SW. This stratum is from 40 to 50 feet wide, and the disseminated minerals are rather irregularly distributed through it but on the tunnel level do not extend north of a quartz vein that has a general northwesterly course and vertical dip. The quartz vein is from a few inches to 4 feet wide, with an average of 15 inches. It contains chalcopyrite, pyrite, and a little galena but in itself is of no economic importance. The shoot of disseminated minerals is about 250 feet long on the 1,200-foot level and about the same length on the 800-foot level. Most of the ore is a fine-grained gray quartzite dappled with sparsely distributed specks of chalcocite and chalcopyrite, Under the microscope the quartzite is seen to be composed dominantly of subangular quartz grains, but there are also feldspar grains and a few shreds of sericite. Chalcocite and chalcopyrite occur for the most part interstitially as thin filaments, but they also replace the quartz and feldspar grains. The disseminated chalcocite and chalcopyrite was noted on the edges of grains of chalcocite and nowhere is chalcopyrite replaced by chalcocite. In places, particularly along the footwall of the ore shoot, the quartzite is fractured and here occurs the richest ore, which consists of chalcocite and chalcopyrite. The chalcocite of this ore is clearly a secondary product replacing chalcopyrite. Oxidation extends in places to the 1,200-foot level on the footwall of the ore body. The ore is of much lower grade than the Snowstorm ore, and only under high prices for copper is profitable extraction possible.

OTHER DEPOSITS.

Other occurrences of disseminated copper minerals in the quartzite stratum of the Snowstorm and National mines are in the Snowshoe and Missoula copper prospects. In the Snowshoe the copper stratum, about 30 feet wide, contains sparsely disseminated pyrite, chalcopyrite, and chalcocite for about 400 feet on the strike of the bed, but the minerals are not abundant enough to constitute ore.

In the Missoula copper prospect, as exposed on the main tunnel level, sparsely and irregularly disseminated copper minerals occur at two horizons in the Revett formation. At one locality 550 feet from the portal these minerals are very scant in amount and no development work has been directed on them, but at 1,250 feet from the portal a northwesterly quartz vein has been explored for many hundred feet, and a shaft on it extends to the surface. The vein is from a few inches to 2 feet wide and in places is mineralized with chalcopyrite and galena. The quartzite strata which the vein cuts contain irregularly disseminated copper minerals, but these are notably more abundant and extend farthest back from the vein, 30 feet at a maximum, where the mineralization in the vein is strongest. Where the vein is barren the quartzite beds contain little or no disseminated copper.

The Copper King prospect, northwest of the National, is developed by a tunnel about 5,000 feet long which cuts rocks of the Newland and St. Regis formations. The principal ore body observed is a northwesterly vein whose average thickness is 1 foot. It contains galena, sphalerite, chalcopyrite, and pyrite in a gangue of quartz and calcite. The wall rocks are slightly mineralized. Another ore body reported to occur south of this vein consists of a band of country rock strongly impregnated with chalcopyrite and galena and is said to be 2 feet wide.

COPPER IN LEAD-SILVER VEINS.

In the zone of lead-silver veins between Wallace and Kellogg, known as the " dry belt," the copper is saved as a small by-product. Of these veins the Caledonia, at the west end of the zone, produced about 600,000 pounds of copper in 1915, about 750,000 pounds in 1916, 550,000 in 1917, 95,000 in 1918, and 55,000 in 1919. Among other mines in this zone in which copper occurs are the Big Creek, Yankee Boy, Polaris, Silverado, and Argentine. In the Caledonia lode chalcopyrite and tetrahedrite were the principal copper minerals; the tetrahedrite carries the high silver content for which the mine is noted. In the other mines of the belt tetrahedrite is the principal copper mineral. One shipment of 74 tons from the Big Creek mine contained an average of 265 ounces of silver to the ton and 5 per cent of copper.

The ore minerals of the lodes are galena, pyrite, tetrahedrite, and chalcopyrite in a gangue of quartz and siderite. Here, as in most of the Coeur d'Alene veins, siderite was first deposited in the veins and later replaced by quartz and sulphides. The copper minerals, tetrahedrite and chalcopyrite, as shown in the Caledonia mine, may have been deposited synchronously, and, at least in considerable part, both are younger than sphalerite and older than galena.

The Caledonia vein occurs in the Burke quartzite; the Big Creek vein is in the Revett quartzite on the hanging-wall side of the Alhambra fault; and the Yankee Boy, Polaris, and Argentine mines are contained in zones of intense shearing, principally in slate of the Newland formation, but also in smaller part in the St. Regis formation. These shear zones strike in a general easterly direction and have steep southerly dips.

CONTACT COPPER DEPOSITS.

A deposit exploited by the Copper Prince Mining Co., near the mouth of Black Prince Creek, on the Chicago, Milwaukee & St. Paul Railway, is of contact origin.[70] It occurs at the contact of monzonite and metamorphosed rocks of the Newland formation, is veinlike in form, and trends about N. 70° W. The chief minerals are pyrite and chalcopyrite, with siderite, quartz, and calcite. These minerals occur in small fissures and joints and as little irregular bunches replacing the country rock. No shipments of ore have been made from the property.

The Copper Prince is the only known contact deposit in Shoshone County that is valuable chiefly for copper, although in the Success deposit, described on pages 95–98, small amounts of chalcopyrite were noted in the predominant zinc and lead ore.

BRECCIATED ZONES IN DIABASE.

Because of the close association of brecciated zones in diabase and siderite veins carrying copper, deposits of this type are described under the heading "Siderite veins" (pp. 110–115).

A theoretical consideration of interest in connection with these deposits is the possible derivation of copper from the magma which gave off the diabase suggested by the abundance of copper-bearing siderite veins in the general area of the Wishards sill. On the other hand, the siderite veins in the northern part of the county are far removed from diabase intrusions. Furthermore, the Wishards sill is probably the oldest igneous rock in the area, as it has undergone great deformation both by folding and by faulting. Although no contact relations between it and the granitic masses were noted, it is undoubtedly of much greater age than the monzonite to which the lead-zinc deposits of Ninemile and Canyon creeks are genetically

[70] Pardee, J. T., U. S. Geol. Survey Bull. 470, pp. 57–58, 1911.

related. As the faulting is younger than the diabase and as the siderite veins of the southern area are related to this faulting or still younger, it is believed that the copper mineralization was not related to the diabase but was a phase of the general mineralization throughout the region and was genetically dependent upon the monzonite magma.

RELATION OF THE COPPER DEPOSITS TO OTHER DEPOSITS IN THE COEUR D'ALENE DISTRICT.

The copper deposits of the Coeur d'Alene district, though of widely different aspect, are believed to be essentially of the same age and to represent an early phase of the zinc-lead deposition. Direct proof that the disseminated copper, that of the siderite veins, and that of the contact deposits were deposited at about the same time is lacking, for in no place, as far as known, do deposits of two of these types intersect. In the Snowstorm, National, and Missoula copper mines, however, the disseminated ore bodies are associated with small quartz veins carrying chalcopyrite and suggest that the disseminated minerals were introduced through these fissures. The copper from each of these types contains appreciable and about equal amounts of gold; concentrates from the Hypotheek, a lead vein in the Prichard formation, contain about $2.50 to the ton in gold; the disseminated ores from the Snowstorm about $2; and the siderite veins as much as $4.

There is no satisfactory evidence as to the age relations of the copper deposition and that of lead and zinc. In a paragenetic study of the lead, zinc, and copper sulphides, where these occur in the same deposit, the primary copper sulphides are found to be older than galena and possibly younger than sphalerite in places, although as a rule the copper minerals preceded those of zinc. This relation is shown in the Caledonia, Bobby Anderson, Black Bear Fraction, and other lodes and in the deposits of the Wardner district and those of Ninemile and Canyon creeks. The order of deposition is siderite, quartz, chalcopyrite, sphalerite, chalcopyrite and tetrahedrite, and galena.

The relative age of the copper and lead-zinc deposits is best determined by a study of the quartz-siderite veins, as discussed on page 62. These veins occur in a broad zone parallel to great west-northwest faults, notably the Osburn and Placer Creek faults. These faults belong to the youngest system of faults in the region, and as shown in the discussion of the Wardner district (pp. 51–55) the ore deposits of that area were formed concurrently with movements on the Osburn fault, probably distributed through a long period of time. The quartz-siderite veins are simple fissures opened during early stages of this faulting movement and filled with siderite, quartz, pyrite, and chalcopyrite. They occur outside the zone affected by later movements and adjustments along the Osburn fault and possibly for that reason were not penetrated by the later zinc and lead solutions. Therefore it would be expected that few siderite veins unaccompanied by zinc or lead minerals would be found in the areas of great disturbance. In the Bunker Hill and Sullivan mine there is an important exception. The Motor seam below its intersection with veins of the Blue Bird type, which contain galena, sphalerite, and pyrite, is a barren siderite-pyrite vein, but above the intersection with the Blue Bird veins it contains large ore bodies that are valuable chiefly for lead.

ANTIMONY DEPOSITS.

Valuable deposits of antimony are known only in the Pine Creek area, although scattered bunches have been found in prospects in the vicinity of Burke, and stibnite occurs as acicular crystals in small vugs in the Golden Hunter lode.

The antimony veins on Pine Creek are widely distributed and apparently are not related to any of the zones of mineralization that characterize this part of the county. The veins have widely diverse dips and strikes. All are contained in Prichard rocks, and their mineral composition is practically identical. The northernmost vein, the Coeur d'Alene Antimony, strikes N. 60° E. and dips 35°–60° NW.; the Albrecht vein strikes N. 10° E. and dips 60° W.; the Pearson strikes N. 40° E. and is vertical; and the Star Antimony strikes east and dips 25°–60° S. Concentrates from the Coeur d'Alene Antimony vein contain as much as $5 a ton in gold, and those from the Star Antimony assay as much as $8 a ton.

Stibnite and quartz are the dominant minerals of the veins, although pyrite is abundantly developed. Galena was not observed, but sphalerite occurs locally. Stibnite replaces the slate and vein quartz and occurs as nests of needles in vugs in the vein matter.

COEUR D'ALENE ANTIMONY.

The Coeur d'Alene Antimony Mining Co. owns two unpatented claims in the SE. ¼ sec. 6, T. 48 N., R. 2 E. The property has been worked intermittently since 1885. The last work was begun in the summer of 1915, and considerable ore has been shipped during the last operations. Ore to the value of $50,000 has probably been produced, though the records of the early operations are very incomplete. About 1,000 tons of stibnite ore is said to have been roasted, and the sublimed product was conveyed to a baghouse and collected. This roasting plant burned down and was superseded by a concentration mill having a capacity of 75 tons a day. The product is shipped to the International smelter. The vein crops out on the west side of Pine Creek, and in the early days boulders of stibnite were worked from the creek bed. The vein occupies a fault or shear zone in black shale and slate of the Prichard formation, which trends N. 60° E. and dips 35°–60° NW. It is exposed in surface workings for about 400 feet, and the greatest depth attained is about 170 feet on the dip of the vein. The footwall formation appears to conform with the strike and dip of the vein for the most part, but the hanging wall is extremely folded and contorted and shows black gouge seams, which in places extend into the ore. The ore ranges from less than 1 foot to 5 feet in width. It consists of stibnite with a quartz gangue. Movement along the fault fissure subsequent to vein formation has greatly crushed the quartz, and much of the stibnite is ground to a slickensided black fine-grained aggregate. The antimony ore is mined southward to a quartz vein that follows a fault plane which trends N. 15° W. and is vertical. The antimony vein is displaced by this fault, but the amount of displacement is not known, as the continuation of the antimony vein has not been found south of the quartz vein. The ore shoot is said to pitch to the southwest. The ore as it comes to the mill is roughly sorted by hand, the high-grade coarse material being picked out. It is crushed and passes to jigs, where three products are made; the first and second grades contain from 50 to 60 per cent of antimony and are shipped; the third product is reground and treated on Wilfley tables, but there is still a considerable loss in the tailings. The company contemplates the fine grinding of the concentrates and treatment by flotation. The ratio of concentration is about 7 to 1, and an average of 3 tons of concentrates a day was being produced. Three carloads of concentrates aggregating 80 tons were shipped in June, 1916. According to the smelter returns from a carload of concentrates shipped in April, 1916, 47,326 pounds of ore yielded 36.36 per cent of antimony and 0.24 ounce of gold to the ton. The antimony was sold at the rate of $3.53 a unit and the gold at 75 per cent of the assay value of the ore at $20 per ounce.

STAR ANTIMONY.

The Star Antimony group consists of nine unpatented claims that occupy a ridge between Stewart Creek and the East Fork of Pine Creek. The deposits were discovered in 1914, but no ore shipments were made until 1916, when three carloads of ore were shipped. About 1,500 feet of development work has been done in three tunnels; the vertical range between the upper and the lower is 150 feet. In the lower tunnel drifts on the vein extend for 400 feet, but the ore so far produced has been derived from the upper workings. The face of the lower tunnel is still 170 feet west of the westernmost part of the drift on the ore shoot in the upper tunnel. The vein strikes east and dips 25°–60° S. It is contained in a fault zone in bluish shale and slate of the Prichard formation. There is much gouge material in the fissure, and the quartz is generally brecciated. Stibnite is the dominant mineral of the ore, but in the lower tunnel as far as the vein was explored at the time of examination it occurs only rarely as nests of needle crystals in quartz vugs and cavities. Pyrite is rather abundant in the lower tunnel, with an occasional crystal of sphalerite, but none was noted in the ore shoot of the upper workings. A sample across a 4-foot quartz lens in the lower tunnel is reported to have yielded $8 a ton in gold. In both the middle and upper tunnels the ore has been stoped for about 60 feet. The ore, with an average width of 3 feet, is contained in a fault zone whose walls are from 4 to 8 feet apart. Stibnite replaces both the slate and the vein quartz; a rich streak of crystalline though sheared stibnite, from 1 to 28 inches wide, occupies the center of the ore body, and the poorer-grade ore consists of disseminated stibnite in the slate. The high-grade shipping ore assays 55 per cent of antimony. Some of the lower-grade material was concentrated by hand jigging to a shipping product, but there is said to be 200 tons of 20 per cent ore on the dumps.

PEARSON PROSPECT.

The Pearson prospect is in sec. 24, T. 48 N., R. 1 E., in a small gulch tributary to Ross Fork. The property was not being worked at the time of examination. About 50 tons of antimony ore is said to have been shipped in 1916. The developments consist of a short crosscut tunnel and drifts on the vein of 150 feet, attaining a maximum depth of 50 feet. The vein is contained in a well-defined shear zone several feet wide which trends N. 60° E. and is vertical. Thinly bedded blue and green slates are the country rocks. The vein has been stoped to the surface through a distance of 100 feet on the strike. The ore appears to be due to the replacement, in whole or in part, of the sheared slate by stibnite. Quartz was not observed, but calcite crystals occur in vugs.

HANNIBAL GROUP.

The Hannibal group of three claims was located in 1914. It is about 1,800 feet south of the Pearson prospect. The developments include a crosscut of 100 feet and a drift on the vein of 150 feet. The vein trends N. 10° E. and dips 60° W. It occupies a shear zone with walls 3 feet apart in slate and a sheared greenish rock that is evidently an altered dike rock on the belt which extends between Ross Fork and the Tiberius prospect. Stibnite occurs in the shear zone in bunches and lenses in places 6 inches thick, but no ore body of workable size has been developed. Several tons of high-grade ore is in the bins, but no shipments have been made from the property.

A few hundred feet west of the antimony vein an inclined shaft has been sunk 60 feet in a shear zone that trends N. 45° W. and dips 60° SW. The sheared rock is 5 feet wide and is sparsely mineralized with galena and sphalerite.

SIDERITE DEPOSITS.

The siderite veins are discussed in connection with copper (pp. 110–115), but because·of their prominence in the region they should be given a place also in a systematic treatment of the deposits.

Siderite veins are extensively developed in the area south of the Osburn fault and in the area north and west of Coeur d'Alene River. In the southern area they are coextensive with the large diabase intrusion known as the Wishards sill, although in many places similar veins occur far removed from it. In the northern area large veins of siderite cross the Little North Fork of Coeur d'Alene River and are locally being exploited for copper.

Copper is nearly everywhere present in the siderite veins of these areas, although in no place has it been found in sufficient quantity to support an extensive mining industry.

The veins range from a few inches to 30 feet or more in width, and some of them crop out for several hundred or even a few thousand feet. The siderite is clearly older than the quartz and chalcopyrite. In the Pine Creek region some of the siderite is notably manganiferous.

Fissure Fillings.

The filling of fissures accompanied by little if any replacement of the wall rock has been of minor importance in the development of the ore deposits of Shoshone County. By far the greater number of the ore bodies occupy spaces much larger than those which directed the ore depositing solutions. Certain of the deposits, however, are closely related to fissure veins, and these are described under this heading. Replacement of the wall rock has been of minor importance in the development of the gold and tungsten bearing veins of the Mullan area. Among the lead-silver deposits only the Jersey "fissures" of the Wardner area have a veinlike aspect, although here it seems almost certain that replacement has played a more important part than the filling of preexisting spaces, and the deposits are considered in another section (pp. 58–60). The Blue Bird "seams" because of a banded arrangement of minerals, appear to have filled open spaces in considerable part. Their commercial value, however, depends more upon their close spacing than upon their size individually, so they are grouped with the other lode deposits (pp. 56–57).

GOLD DEPOSITS.

HISTORY AND PRODUCTION.

Gold was first discovered in Shoshone County on Prichard Creek in 1882, and placer mining began on many claims early in 1884. During the next four years most of the gulches were worked out, and production then largely ceased. At different times dredging has been attempted, but gravels in many places over 30 feet deep and an irregular bedrock of steeply dipping beds of slate and quartzite have hindered successful operation. Recently the extensive gravel bed along Prichard Creek from Raven to its mouth, said to comprise about 8,500,000 cubic yards with an estimated gold content of more than $1,350,000,[71] has been carefully drilled, and a dredge is now operating successfully on it. Gold quartz mining began shortly after the placer mining but soon declined, though somewhat more slowly, to an even lower level of annual production.

Of the total production of gold in the county of $6,155,331 to the end of 1920, probably less than $2,000,000 has come from gold deposits, the balance being derived from the small proportion of gold in lead and copper ores. About three-fourths of the production from the gold deposits was made during the four years after the beginning of mining. In 1914 nine placers produced $2,060 in gold and 15 ounces in silver, and no production from the gold veins was reported; in 1920 the placer gold production was $95,775.

DISTRIBUTION.

The principal gold deposits are in the vicinity of Murray. Smaller deposits occur north and east of Kellogg and in the St. Joe area. The mines near Murray include the Yosemite, Daddy, Mother Lode, Treasure Box, and Occident, on Ophir Mountain, from 1 to 2 miles southeast of the town. To the north, across Prichard Creek, is the Golden Chest, and west of this, on Gold Run Creek, are the Gold Back and Pilot. The Golden Winnie and Buckeye Boy are west of Murray, and the Tiger group south of Murray. East of Kellogg the Wisconsin, New Jersey, Alhambra, and Enterprise are the principal gold-bearing veins. In the St. Joe area are the Ward, Big Elk, and Silica properties. Gold also occurs in several of the lodes on Pine Creek, being an important accessory element in the ores of the Star Antimony, Hypotheek, and Coeur d'Alene Antimony mines.

GEOLOGIC RELATIONS AND STRUCTURAL FEATURES.

The deposits near Murray and those near Kellogg, except the Alhambra gold vein, are in the Prichard formation. The veins in the

[71] Yukon Gold Co. Ann. Rept., 1916.

St. Joe area and the Alhambra are in the Newland ("Wallace") formation. Most of those about Murray lie in large part parallel to the bedding of the inclosing formation, but elsewhere they cut across the strata irrespective of their attitude. The veins near Murray, except the Golden Chest, described under tungsten deposits, have not been worked since Ransome's examination in 1904. His description in part follows:[72]

The veins in some places occur singly, but more commonly they are in groups, individual veins being separated by a few inches or a few feet of slaty country rock. An imbricated arrangement is common, one vein gradually pinching out while an overlapping vein between adjacent beds becomes correspondingly wider and thicker. Although some single veins may persist for hundreds of feet without cutting across the planes of stratification, such crossings may be observed here and there, and crosscutting stringers of quartz linking neighboring bed veins are fairly abundant.

The veins as a rule are narrow and can be stoped only by the removal of considerable waste rock. Although a maximum width of about 10 feet is attained by one of the veins in the Golden Chest mine, the average width of all those that have been worked is probably nearer a foot. The contact between quartz and slate is sharp and usually unaccompanied by gouge. * * *

Where the beds are thrown into sharp local folds the bed veins are also folded, as may be seen on the Martin level of the Golden Chest mine and in the Crown Point mine. In both of these mines also the veins are dislocated by faults. * * *

The only important fissure vein which cuts directly across the bedding of the slates [in the Murray area] is the Mead vein, on Ophir Mountain. This strikes N. 17° E. and dips 75° E. It is accompanied by two parallel bed veins, which have furnished the ore to the Mother Lode, Treasure Box, and Occident mines. These bed veins lie east of the Mead vein and dip to the northwest at angles of about 18°. Their relation to the Mead is not shown in the underground workings, and it is not known whether all three veins are of the same age or whether the Mead vein occupies a fault fissure which has cut the bed veins. The latter, so far as known, do not cross the Mead to the west.

A conspicuous banding, much of it very irregular, is highly characteristic of the bed veins of the Murray district but is not found in those veins or stringers which clearly cut the planes of bedding. * * * As a rule, the narrower the vein the more regular and noticeable is this structure. In the wider veins the middle portion may be massive quartz and the banding may be confined to the parts of the vein near the walls. The banded appearance is due to the alternation of thin plates of quartz, usually a fraction of an inch in thickness, with much thinner sheets of dark material. In some places this dark material is a film of sheared slate, such as may be seen adhering to the side of the vein when a part of the wall rock is stripped away; but in many places the dark bands are due mainly to the presence of minute crystals of pyrite and other sulphides closely aggregated along parallel planes in the quartz.

The explanation of the banding which appears most in harmony with the facts is that the veins have been formed by successive enlargements of the fissure, whereby the quartz previously deposited was separated from the wall with a film of slate adhering to the side of the vein. * * *

[72] Ransome, F. L., and Calkins, F. C., The geology and ore deposits of the Coeur d'Alene district, Idaho: U. S. Geol. Survey Prof. Paper 62, pp. 141–142, 1908.

The microscope shows that in addition to the banding just described some of the quartz has a true ribbon structure, being traversed by microscopic cracks generally parallel with the .walls of the vein. These cracks have a minute suture-like irregularity and cross the grains of quartz with little or no faulting. They are themselves here and there faulted, however, by other microscopic cracks transverse to the vein. Both sets of fractures, particularly the older, have afforded passage to solutions which have deposited small crystals of pyrite along them. These crystals almost invariably have diameters greater than the widths of the cracks and the pyrite has grown by replacement of the adjacent quartz. This ribbon structure is much less conspicuous than the banding due .to successive reopenings of the vein fissures.

A large nearly vertical vein not described by Ransome is known as the Gold Back. It appears to occupy the Murray Peak fault zone and is said to be traceable from the mouth of Niagara Gulch southward nearly to Tiger Peak. It was examined in Ophir Gulch, near the mouth of Gold Run, and in Oregon Gulch. The vein consists of quartz and much pyrite and pyrrhotite, mixed with slate, together with irregularly distributed chalcopyrite, sphalerite, and galena. Near the mouth of Gold Run the vein, or more accurately the ledge, is about 30 feet wide, and in the other two localities about 10 feet wide. Wherever assayed it is said to have yielded about 80 cents to the ton in gold. Mr. John Doctor, owner of the Gold Run group, is authority for the statement that it definitely cuts the Mother Lode bed vein.

The veins east of Kellogg consist of white and bluish-gray quartz, range from seams to veins 12 feet or more wide, and strike northwest and dip southwest. They are fissure fillings, in most places not banded, and contain sulphides and locally many slate fragments irregularly scattered through the quartz. Most of them have been fractured, and considerable quantities of the sulphides occupied the openings thus formed, although a greater amount does not appear to be related to openings of any sort and probably was formed contemporaneously with the gangue material. Several of the veins contain a little siderite intermixed with the quartz, and it is probable that most of the sulphide is in siderite areas. In general the walls are well defined and almost free of selvage. The veins are described in some detail by Hershey.[73]

The gold-bearing veins of the St. Joe area described by Pardee[74] comprise several quartz fissures inclosed in the Newland formation. The Ward, Big Elk, and Silica groups include the principal deposits. The Ward, on a southeast spur of Wards Peak, contains a mineralized zone about 50 feet wide in which are numerous seams and veins of quartz, calcite, and siderite carrying chalcopyrite, pyrite,

[73] Hershey, O. H., Origin and distribution of ore in the Coeur d'Alene (private publication), pp. 6–9, 1916.
[74] Pardee, J. T., Geology and mineralization of the upper St. Joe River basin, Idaho: U. S. Geol. Survey Bull. 470, pp. 46–47, 1911.

and chalcocite and assaying about $4 to the ton in gold. The Big Elk, also on a shear zone, is about 6 feet wide and similar mineralogically, but the sulphides are in bunches, and the better parts assay about $10 in gold and 2 or 3 ounces in silver to the ton. The Silica is on three parallel veins, the largest of which is a fault zone 6 feet wide consisting of crushed rock and angular fragments of quartz. Pyrite and chalcopyrite, containing from $4 to $12 to the ton in gold, occur locally.

THE ORES.

The ore of the Murray veins consists of the banded quartz with variable amounts of pyrite, chalcopyrite, galena, sphalerite, and free gold. Siderite has not been observed. Scheelite is irregularly distributed in the Golden Chest and Golden Winnie, and from its reported occurrence in the old placer deposits it is probably present in several other veins where it has not been mined.

Much of the surface ore is said to have been very rich in gold compared to that of the deeper levels, nearly all of which are opened by tunnels. In the Daddy and Mother Lode mines ore from the upper levels is said to have averaged $30 to $50 to the ton, much of it in visible gold, whereas that obtained 100 to 300 feet lower in the vein assays from $2 to $7. The upper levels of the Golden Chest are said to have produced ore containing much wire gold and averaging above $50 to the ton, whereas ore mined at greater depth has run from $7 to $10 to the ton in gold. It is probable that similar conditions exist in the other veins, for had not the upper parts of them contained much richer ore than the deeper parts later opened, mining with the crude equipment available would not have been pursued so generally at the time they were worked. The obvious inference is that the gold veins were enriched to an average depth of 50 to 100 feet and locally, as in the Golden Chest mine, to a depth of possibly 300 feet.

The unaltered ore consists of variable amounts of pyrite, sphalerite, galena, arsenopyrite, chalcopyrite, and rarely magnetite and pyrrhotite in a gangue of white quartz. Most of the gold is said by local assayers to be in the pyrite, concentrates of it having yielded assays of over 100 ounces to the ton. The oxidized ore, specimens of which were seen in the collection of Mr. Adam Aulbach, of Murray, consists in general of a slightly honeycombed iron-stained quartz with free gold along fractures and less commonly as grains in cavities from which sulphides have been leached. Manganese, which is known to be favorable to the migration of gold in the oxidation of mineral deposits, is not abundant, although the dendritic form was observed in some of the specimens examined.

The veins near Kellogg are similar mineralogically, although chalcopyrite is much more prominent and ore from some of them has been more valuable for copper than for gold.

RELATION OF THE GOLD VEINS TO THE OTHER ORE DEPOSITS.

It is noteworthy that the gold veins contain the same minerals as the lead-silver lodes but in vastly different proportions. Deposits which may be considered transitional between the two types, however, are not known with the possible exception of the little-developed Gold Back vein, which has been mentioned under both groups. The Terrible Edith vein, a lead-zinc deposit about 1½ miles north of the Golden Chest mine, contains about $2 to the ton in gold in material hand sorted for shipment. It does not, however, resemble the white quartz gold veins in general aspect. Certainly no gradation exists between the gold veins and the lead or zinc lodes such as exists between the lead and zinc lodes, where several deposits contain about equal amounts of each metal and the lodes present the same minerals, formed in the same order and distributed in the same manner.

Between the gold veins and the copper deposits, on the other hand, gradations may be recognized, as several of the small veins near Kellogg and in the St. Joe area may be considered equally valuable for each of the two metals, and gold was an important constituent of the Snowstorm ores. Between the copper and the lead-silver deposits gradations may be recognized, as pointed out elsewhere (pp. 118–119).

TUNGSTEN DEPOSITS.

Scheelite, one of the more common tungsten minerals, has been known to occur in the gold veins near Murray for a number of years, but not until the period of high prices in 1915 were any of the mines worked exclusively for it. Two mines in the district have shipped tungsten ore since that time—the Golden Chest and the Friday, formerly the Golden Winnie. Prior to July, 1916, the Golden Chest had made 11 shipments, averaging about 65 per cent of tungsten trioxide, the last shipment containing 2 tons, and the Friday had shipped 6 tons in all of 60 to 70 per cent ore. A little ore has also been taken from outcrops on the west side of Ophir Gulch. Specimens of scheelite have been found along the lower part of Idaho Gulch, the upper part of Trail Creek, and the upper part of Pony Gulch.

In the Golden Chest mine massive scheelite forms lenses and pipes inclosed in quartz. These are confined to the Klondike shoot, an enlargement of the bed vein made up of massive quartz in marked contrast to the ribbon quartz characteristic of narrower portions of the vein. Recent development work has been done from the Idaho tunnel, which enters from the main valley and is at a level 231 feet lower than the Pettit tunnel, the lowest adit at the time of the earlier examination. The distribution of the scheelite was well shown in July, 1916, in No. 1 raise, about 30 feet above the Idaho level.

Here a circular pipe of nearly pure pale-brown scheelite had been followed for about 10 feet. At the point of discovery this pipe was very small, but within a short distance it enlarged to 2 feet in diameter and beyond gradually diminished to 10 inches in the face seen. In another face four pipes, somewhat irregular in cross section, ranged from 4 to 6 inches in diameter. In two rounds these had disappeared, but others were picked up beyond.

The scheelite crumbles readily, and in mining much of it is caught directly in powder boxes; the remainder is recovered by hand jigging after the coarser waste is picked out. The scheelite and gold occur principally in separate parts of the vein, so that ores of the two metals may be broken separately. The jig product, however, carries a little gold. This occurrence of the tungsten mineral seems to be somewhat different from that higher in the shoot, as described by Ransome,[75] who states that the scheelite " is intimately associated with the quartz and like that mineral incloses the various sulphides found in the vein. * * * Whether the sulphides embedded in it contain as much gold as those in the quartz is not known."

The Friday (Golden Winnie) deposit is a ribbon quartz vein which dips 20°–45° N. and is offset as much as 73 feet by cross faults. The scheelite occurs in small bunches and scattered seams a few feet long in the white ribbon quartz. The ore as concentrated in a small mill contains about $80 in gold to each ton of concentrate which averages 65 per cent tungsten of trioxide. The 6 tons shipped has been taken out at intervals along 700 feet of vein developed near the creek level.

The prospect on the west side of Ophir Gulch near the junction of its main branches is developed by several shallow open cuts along a general N. 50° E. course. Some of these cuts expose quartz seams as much as 6 inches wide which contain well-formed crystals of scheelite and a little galena irregularly distributed. In one open cut the quartz is about 4 feet wide and is traversed by two distinct seams of scheelite each about three-fourths of an inch thick.

A series of metallurgic tests were run on the tungsten ores by Goodrich and Holden in 1916. Their results suggest treatment by gravity concentration followed by roasting and magnetic separation to remove the pyrite,[76] but it is by no means certain that enough tungsten exists in the deposits to suport an industry under normal market conditions.

GRADATION BETWEEN THE SEVERAL GROUPS OF DEPOSITS.

The foregoing grouping of the deposits, adopted in order to emphasize differences in genesis, wall rock, form, and substance, may be

---

[75] Op. cit., p. 146.

[76] Goodrich, R. R., and Holden, N. E., Experiments in the recovery of tungsten and gold in the Murray district, Idaho: Am. Inst. Min. Eng. Trans., vol. 58, pp. 224–231, 1917.

misleading unless its arbitrary nature is emphasized. With the possible exception of the gold-quartz veins near Murray, which in general aspect stand rather distinctly apart from the other deposits, all transitions may be traced between such widely diverse types as the disseminated copper of the Snowstorm mine, the chalcopyrite-siderite lodes of the southern copper area, the Success deposit with contact-metamorphic characteristics, the zinc lode of the Interstate-Callahan mine, and the Bunker Hill and Sullivan galena-siderite deposit. Even the antimony lodes are linked to the others by many essential features in common.

Within the Success deposit, which is principally inclosed in slate but locally in monzonite, transition occurs from ore in which contact-metamorphic silicates are intergrown with sulphides to zinc ore like that in the west end of the Morning lode, where none are found, and thence to the zinc-bearing lead-silver ore of the eastern part of this lode. Both deposits have considerable magnetite, but siderite, which is abundant in the Morning, is almost entirely absent from the Success. It occurs, however, locally in the Rex, a deposit with many features typical of contact metamorphism. Pyrrhotite is widely distributed in the zinc lodes, such as the Rex and Highland Surprise, but also occurs in such lead-silver deposits as the Tiger-Poorman and Standard-Mammoth. Some of the zinc mines contain more quartz than siderite, a feature equally conspicuous in the Hecla and Jersey lodes, typical lead-silver deposits. Indeed, between such extreme types as the Interstate-Callahan zinc lode and the great March shoot of lead-silver ore in the Bunker Hill and Sullivan mine, occur deposits representing every gradation in the ratio of lead and zinc, composition of wall rock, form of deposit, and relative abundance of the gangue minerals. In all the deposits the same minerals are present, deposited in the same sequence, the only difference being in their relative abundance.

Chalcopyrite, the predominant primary copper mineral of the Snowstorm disseminated deposit, is present as an accessory constituent in nearly all the lodes of the area. In the zinc-lead ore of the Black Bear Fraction it occurs locally in recoverable quantities. In lode form chalcopyrite is widely distributed in the upper St. Joe River basin and constitutes an ore of copper at the Empire mine on the Little North Fork of Coeur d'Alene River. Gold is nearly everywhere associated with it, and the principal gangue mineral is either siderite or quartz. From these lodes there is complete mineralogic gradation, represented by many deposits, to the typical gold veins of the Murray district, in which all the basic sulphides found in the other deposits occur in minor amounts. The difference lies in the general aspect of the Murray veins, which may or may not be of important significance. The gold veins lie parallel to the bedding and are strikingly banded by numerous thin partings of slate which parallel the walls. Where they cut across the beds, as locally in the Mother Lode mine, these partings are absent, and here the veins are in general appearance quite like such deposits as the New Jersey gold-quartz vein, near Kellogg, which seems to be merely a variation from the Wisconsin and Teddy copper-gold veins situated near by. The writers favor the view that the unique aspect of the gold veins is due to a lack of fissuring in the Murray area at the time of ore deposition as compared with other mineralized parts of the county, although the explanation is recognized as not fully satisfactory.

Barite is a gangue constituent in the Gold Hunter lead-silver lode, the Morning lead-silver-zinc lode, the Carney copper, sparingly in the Bitter Root galena-siderite lode, and rarely in the Senator Stewart mine. More than other minerals of the county it seems to have a narrow geographic distribution.

Stibnite, the source of antimony in Shoshone County, is widely distributed. It occurs as a rare constituent of many of the lead-silver ores, is associated with zinc-lead ores in the Pine Creek district, and is found as fairly clean lenses in the Star Antimony and Coeur d'Alene Antimony mines, on Pine Creek, and in the Stanley mine, near Burke. The antimonial mineral tetrahedrite also occurs widely distributed in the lead-silver lodes and serves as a tie between the antimony deposits and the copper deposits.

STAGES OF MINERALIZATION.

The metalliferous deposits of Shoshone County are believed to have all been formed during one general period of mineralization. That this period was characterized by structural disturbances is clearly shown by the prevailing occurrence of galena and other late minerals as seams in the older minerals and by the relations in the Wardner area, where there are three systems of lodes with different strikes and dips, all of which contain the same essential minerals but in different relative amounts. A sequence of deposition from siderite accompanied and followed by pyrite successively to quartz, sphalerite, and galena is characteristic of all the lodes. (See p. 135.) It would seem, therefore, that movement concurrent with deposition, so that lodes of different systems were reopened at different times, is a sufficient explanation of preponderant galena and siderite in the Bunker Hill area, quartz and galena in the Jersey area, a mixture of the four predominant minerals of the region in the Blue Bird lodes, quartz and galena in the Hecla, sphalerite, quartz, and siderite in the Interstate-Callahan, and quartz and pyrite in the gold area near Murray.

The principal suggestion of more than one period of mineralization, each characterized by the same minerals formed in the same order but in different relative amounts, is the offset of 800 feet in certain of the Jersey fissures on opposite sides of the Bunker Hill lode. It is believed, however, that this offset is due to movement on the Cate faults in a direction parallel or nearly parallel to the pitch of the March ore shoot. The Lower Cate fault cuts the March shoot in many places, as illustrated on Plate XII, and the marked curvature of the fault plane over the ore shoot makes necessary an assumption of agreement in the direction of movement and the pitch of the shoot. As the Jersey veins dip southeast, a movement downward and westward along the pitch of the March shoot may fully explain the observed horizontal offset of 800 feet to the west, south of the Upper and Lower Cate faults. Drag phenomena on the faulted ends of the Jersey veins favor this view. It is not impossible, however, that the observed offset may be due in large part to premineral movement and that the fissure segments, having a common strike and dip, responded synchronously to dynamic forces during mineralization. The relation of the same two veins to the Last Chance fault on levels 12 and 13 of the Bunker Hill and Sullivan mine lends some support to such a view. Eastward the veins converge, and at their point of union the filling of each merges with that of the other and continues unbroken beyond with no evidence of any difference in age of the two veins; certainly they are essentially contemporaneous. On levels 12 and 13, however, the Last Chance fault offsets the east vein as much as 100 feet but does not affect the other vein, only 100 feet away. This can not be explained by strike and dip relations, because the two veins are essentially parallel in this part of the mine and the fault is nearly at right angles to them (Pl. XII). The evidence is somewhat discordant, as on level 14 both veins are offset a few feet by what appears to be the Last Chance fault. As the fault plane is mineralized in places, however, it seems clear that it is of premineral age, and the relations on levels 12 and 13 strongly suggest that its principal movement is younger than the east vein fissure and older than the west one.

In view of these alternatives it seems to the writers to be unreasonable to assume a difference in age between the Bunker Hill and the Jersey lodes because of the offset of the latter. Neither is the general observation that the Jersey lodes are more broken by postmineral faults than others in the district necessarily of great significance. They lie athwart the lines of slicing that characterize the local structure, whereas the other lodes are approximately parallel to them.

Although the evidence is believed to support the conclusion that there has been only one general period of ore deposition in the county and that absolute gradation exists between all the types of deposits with the possible exception of the gold veins near Murray, there is undoubtedly a geographic difference in the ore deposits. Barite as an essential gangue mineral occurs only in the Gold Hunter lode and the east end of the Morning lode. Stibnite is found in commercial quantity only in the Pine Creek district, and here also the siderite appears to be more highly manganiferous than elsewhere. The gold-quartz veins occur principally in a part of the county where lead and zinc mines, although present, are not highly productive. Similarly the siderite-chalcopyrite deposits are localized in two general areas apart from large lead and zinc deposits. These general facts, however, are considered as evidence of regional variations in the mineral-depositing solutions rather than of distinct periods of mineralization.

It remains to inquire whether the ore deposition as a whole was a continuous or an interrupted process, even though only one general period of mineralization is recognized. May all the variations be accounted for by the temperature cycle at any given point in the lodes—a rising temperature as the solutions advanced and a falling temperature as the activity died down? The answer is not clear. It is noteworthy, perhaps, that siderite veins almost devoid of other minerals are extensively developed in the area and that a few barren quartz veins occur. These veins would seem to favor the view that mineralization proceeded by more or less distinct pulses and that openings existing during one pulse were closed during others. This would account for the Motor vein, which is a siderite-pyrite mass except above its intersection with narrow veins rich in galena. It is supported also by the occurrence of lime silicates intergrown with galena and magnetite locally in the Helena-Frisco mine, but possibly it does not account for the marked increase of zinc westward in the Morning lode.

The view favored by the writers, although not fully proved, is that the process was continuous and that it varied in intensity from time to time. Most of the apparently sharp breaks in deposition are doubtless to be accounted for by structural readjustments concurrent with mineral deposition, so that now one and then another set of fractures served as channels for the ore-depositing solutions. If solutions undergoing gradual change as deposition proceeded are assumed, and this assumption seems abundantly warranted from the paragenesis of the ore minerals, a recurrent opening of fractures in different directions at different times as the process proceeded affords a satisfactory basis for interpreting all the observed differences in the deposits.

PARAGENESIS OF THE PRIMARY ORE MINERALS.

The outstanding feature of the relative age of the ore minerals in the deposits of Shoshone County is a fairly definite sequence, readily recognized in most of the lodes, of the four principal minerals—siderite, quartz, sphalerite, and galena, named from earliest to latest. Locally the recurrence of conditions favorable to the deposition of one or another caused repetition within the series, but the relative amount of any one of them occurring as a second or a third generation is so small as to be almost negligible. In some places, particularly where intersecting veinlets are absent, contemporaneity of deposition of two or more of them is suggested, but in the great majority of places where the evidence is clear the same order of development is shown. During the examination in 1916 very few stopes were seen in which galena was not obviously the last mineral deposited. Its characteristic mode of occurrence throughout the lodes is as reticulating veinlets and irregular bunches due to replacement from them. Even where solid masses of galena occur this relation is characteristically developed along their borders. The evidence of a definite sequence is particularly clear where the end members of the series are both strongly developed. Thus in the March ore shoot of the Bunker Hill and Sullivan mine the ore consists of siderite and galena with only traces of quartz and sphalerite. Here the ore has retained its characteristics from the surface to the deepest level, more than 4,000 feet lower on the pitch of the ore body, so that Ransome's description, written in 1908, fully applies to ore of the present day. He says:[77]

In the richest ore the galena predominates and forms irregular and ill-defined bunches. These grade into massive siderite in which the galena forms countless small reticulating veinlets. Although many of the veinlets are of microscopic width, the absence of definite walls to the little fissures is noticeable, and it is evident, even without the use of the microscope, that the galena has not merely filled the cracks but has been deposited metasomatically at the expense of the siderite. At the intersections of the cracks the galena has gathered in little bunches. In a more advanced stage of replacement the veinlets are wider and more irregular, the bunches of galena at the intersection are larger, and the final stage of the process is a mass of nearly solid galena with perhaps here and there a shadowy remnant of siderite.

The disposition of galena in siderite in the March ore shoot is perhaps sufficient proof that much of the iron carbonate was deposited and fractured before the principal lead deposition began; but additional evidence may be cited. In the Motor shoot, in the east end of the same mine, a group of vertical seams rich in galena intersects a great siderite mass of low dip at about the 850-foot

---

[77] Op. cit., p. 108.

level. Above the intersection a vast tonnage of lead ore, consisting of galena in bunches and reticulating seams in siderite, has been mined, but below this level the ledge, although of equal size, consists only of siderite and associated pyrite.

In the Morning mine the same relation of galena to siderite occurs as is found in the March ore shoot. Indeed, this relationship of the two minerals is characteristic throughout the ores of the area, as would be expected if they represent the end members of a fairly definite sequence of deposition. Where two minerals close together in the sequence occur in juxtaposition the evidence of their relative age is not so conclusive and is locally contradictory. This statement applies principally to quartz, which certainly had a long period of deposition. Throughout the region, however, with the few exceptions described later, quartz is younger than siderite. In describing the " southern copper area " Calkins and Jones [78] say:

> An interesting fact of paragenesis observed in many prospects of this area is the clear priority of siderite to quartz. This is suggested by the presence of many quartz veinlets cutting siderite. It is strikingly demonstrated by the occurrence of abundant perfect siderite crystals, commonly 1 or 2 inches in diameter, partly embedded in quartz which has evidently filled cavities lined with druses of the carbonate.

Quartz attains its maximum development in the Jersey fissure zone, in certain of the Canyon Creek lodes, and in the gold-bearing veins. It is believed to have had a longer period of deposition than any other of the ore minerals. In the Jersey ores it clearly follows zones along which siderite had earlier replaced quartzite and is itself replaced by sphalerite and galena. In the Standard-Mammoth ore body, on the other hand, two generations of quartz are clearly recognizable—one older than sphalerite and galena and the other traversing them in veinlets having clean-cut walls. In the Black Bear deposit quartz is replaced by galena with a very sparse development of all the other minerals, and in the lower levels of the Hecla mine quartz greatly exceeds siderite. Here the quartz is dominantly older than the sulphides, but in many places seams and stringers of it traverse the lode and extend into the inclosing quartzite. In the Helena-Frisco mine quartz is the oldest mineral now in the lode and is replaced by sphalerite, which is traversed by galena. Locally it is replaced by magnetite and galena intimately intergrown and containing felted aggregates of actinolite. In the gold veins quartz is the dominant gangue mineral and in part is older than the sulphides but in the main seems to be nearly contemporaneous with them.

Although the age relations of the four principal minerals of the lodes are clear and conclusive for the deposits as a whole, the position of the accessory constituents in the general sequence can not be stated

[78] U. S. Geol. Survey Bull. 540, pp. 195–196, 1914.

with so much assurance. Pyrite in the main accompanied or closely followed siderite. In the Bunker Hill and Sullivan mine it occurs as disseminated grains widely distributed in the quartzite and in distinct veinlets, in addition to being an abundant element in the Blue Bird ores. Some of the veinlets cut quartzite containing disseminated siderite, indicating that here the iron sulphide followed the iron carbonate. In the Phil Sheridan tunnel, however, a large lode is made up almost exclusively of siderite and pyrite so intimately associated that no difference in age could be made out. Here one specimen may afford evidence that siderite is the younger, and another one obtained near by may show it to be clearly the older. In the Canyon Creek lodes pyrite is nearly everywhere intimately associated with siderite. Seams of pyrite locally cutting all other minerals except galena suggest, however, that it had a comparatively long period of formation.

The magnetic sulphide pyrrhotite was one of the earlier minerals formed but is not an important constituent of the ores except in a few places. Closely related to it in age is magnetite, which occurs most abundantly in the Frisco lode. Here, however, some of the magnetite was developed late in the general period of mineralization, as shown by its intimate intergrowth with galena and felted aggregates of actinolite in the west end of the mine on the 1,800-foot level. Near this stope a veinlet of vesuvianite cuts monzonite, through which pyrite and a little sphalerite are disseminated. Magnetite found elsewhere in the lode is cut by sphalerite, and this in turn by galena, so that the relations seem to indicate a later pulse of mineralization characterized by higher temperature and more complex solutions than those which accomplished most of the deposition.

Tetrahedrite and chalcopyrite were probably formed at about the same time, although the evidence is not conclusive. In the Caledonia lode seams of galena traverse both minerals and both traverse sphalerite, but nowhere was one of them seen to cut the other. Their characteristic mode of occurrence here is as rounded kernels, principally of microscopic size, scattered through galena, although in the central part of the main ore shoot massive forms are not uncommon. In the Hypotheek similar kernels are prevalent, but veinlets of tetrahedrite cutting quartz are common. Here the tetrahedrite, at least in part, is later than the chalcopyrite, and the chalcopyrite and some of the tetrahedrite are older than the galena, but the relative age of the galena and gray copper veinlets was not observed. In the Hypotheek ores arsenopyrite is about as abundant as pyrite and appears to have been formed during the same stage of mineralization.

Evidence as to the age relations of barite is not satisfactory, although it is in part younger than quartz and probably all of it is older than galena. In the Hunter deposit, where it is most abundant, openings during the general period of mineralization were formed parallel to the walls, so that intersecting veinlets are rarely available as evidence of paragenesis. In a number of places, however, bands of galena were observed to cut obliquely across bands of barite. That quartz is older than barite in the Hunter lode is indicated by its common occurrence as rounded patches and kernels embedded in the barite. In the Morning lode, on the other hand, wherever any difference in age between barite and quartz could be determined quartz is the younger. It seems likely, therefore, that barite and quartz were formed at about the same time, with conditions favorable to one or the other representing only slight differences.

Stibnite is exceptional in the deposits and seems to have been a late mineral to form. In the Hunter ore it occurs as nests of acicular crystals occupying small cavities in quartz and rarely in siderite.

OXIDATION PRODUCTS AND THEIR RELATIONS.

Oxidized ores have long since ceased to be an important factor in the metal production of Shoshone County, although two of the mines, the Caledonia and Hypotheek, still produce ore of this type. The principal changes in the process of oxidation involve the transformation of argentiferous galena to cerusite and of native silver and manganiferous siderite to limonite and oxides of manganese. Massicot occurs in diminishing amounts down to 850 feet in the Caledonia lode, was observed sparingly in several parts of the Hypotheek mine, and has been reported in subordinate quantity in the oxidized portions of several other deposits. Pyromorphite occurs near the surface in many of the deposits and, as concluded by Ransome,[79] " seems to have been one of the latest minerals to form in the upper part of the zone of oxidation, for it occurs in crusts and implanted crystals upon other products of oxidation such as limonite." Its greatest depth of occurrence seems to have been on the Caledonia 900-foot level, where doubly terminated crystals were found attached to fresh galena. Plattnerite, the dioxide of lead, is reported by Ransome[80] from the You Like vein and by E. V. Shannon[81] as lumps and crusts in carbonate ore from the Mammoth, Last Chance, and Hercules. In the Mammoth lode it occurred sparingly to a depth of 200 feet but elsewhere to shallower depths. Argentite is exceedingly rare in the district, and cerargyrite is not known to occur. Tetrahedrite in the Caledonia lode was considered by Shannon[82] to be a secondary min-

[79] Op. cit., p. 132.
[80] Idem.
[81] Oral communication.
[82] Shannon, E. V., Secondary enrichment in the Caledonia mine, Coeur d'Alene district, Idaho : Econ. Geology, vol. 8, p. 469, 1913.

eral, but no evidence of such origin appears in the material examined by the writers, and certainly most of it is primary. Above the 500-foot Caledonia level it was particularly abundant and "nearly everywhere it covered cores of chalcopyrite," this being the principal evidence for Shannon's conclusion. On the 700-foot level in 1916, however, intersecting veinlets clearly showed tetrahedrite, in part at least, to be older than chalcopyrite, which was older than galena.

Malachite and azurite stains were observed at several places in the Caledonia lode, and malachite occurs in the Hypotheek. Native copper in small scales, threads, and wires along fractures in iron-stained quartz is conspicuous in many places between the 500 and 950 foot levels of the Caledonia mine. At higher levels cuprite and sooty covellite are reported to have been abundant, the former in amounts sufficient "to make it profitable to sort copper ore from the lead ore." [83]

Smithsonite and calamine, which in other districts are commonly present in oxidized zinc ores and oxidized lead ores containing zinc and which are readily identifiable, were not observed by Ransome or during the more recent investigation, although the abundance of lead carbonate affords a reasonable basis for suspecting the presence of the common carbonate of zinc.

Cerusite is the most noteworthy secondary mineral of the deposits and occurs in massive crusts and well-developed crystals instead of in the incoherent "sand carbonate" prevailing in the Texas and Wood River districts of the southern part of the State. Anglesite is very rare in the deposits, whereas elsewhere in the State it is a characteristic intermediate form between galena and cerusite. The carbonate is mixed with iron and manganese oxides, often in a siliceous or jaspery matrix. As a rule such ore is porous and contains many vugs from an inch to 6 or 8 inches across, in marked contrast to the dense primary material. The change of galena to cerusite involves an increase in volume of 28 per cent, and that of siderite to limonite a decrease of 26 per cent. Thus where the volume of siderite exceeds that of galena, oxidation may result in increased porosity without the loss of material. Near its outcrop the Caledonia lode was porous, and lead, silver, and copper were present only in small amounts, but they became successively more abundant at increasing depth, suggesting that a noteworthy transfer of material has occurred and that some of the increase in porosity in many of the lodes is due to leaching. The general coherency of the oxidized lead ore may be due to the absence of anglesite in most places as an intermediate stage in its development. Throughout the district cerusite seems to have developed

---

[83] Shannon, E. V., op. cit., p. 566.

directly fom galena, and anglesite, which is common elsewhere in a boundary zone between the sulphide and carbonate, is characteristically absent. This may account for the absence of "sand carbonate," for in the formation of the carbonate through the sulphate there is a 52 per cent increase in volume followed by a 24 per cent decrease, conditions favoring an incoherent product. Throughout the deposits anglesite is absent or is a mineralogic curiosity. In material collected by the writers it appears only in one specimen from the Hypotheek lode.

The depth of oxidation is irregular. Galena cropped out at the surface of the March ore body, but in the Caledonia lode on the same upland mass oxidation is extensive to the lowest or 950-foot level. According to Ransome,[84] the Morning, Gold Hunter, and Standard Mammoth lodes all contained much sulphide ore at the surface. In the Morning lode the lower limit of oxidation was about 200 feet deep. The Hypotheek lode is oxidized in large part to a depth of 1,100 feet, the lowest level in the mine, or approximately 700 feet below the level of Coeur d'Alene River at a point 2 miles to the north. The lower limit of oxidation is not consistently related to mineral composition or to elevation above adjacent drainage lines, so that local permeability must be considered the chief factor in determining its position. The great depth of oxidation in the Hypotheek lode seems to imply an elevation of the drainage level for the region, as this lode occurs in a broad area of the Prichard formation which failed to reveal any evidence of downfaulting of the part inclosing the lode. The valley of the South Fork of Coeur d'Alene River is an aggraded channel which attained its maximum depth prior to the interruption due to glacial outwash in the vicinity of Coeur d'Alene. Corroboratory evidence of the extent of aggradation subsequent to these events has not been recognized further than that St. Joe River and many lakes in the region occupy drowned valleys. It seems likely, however, that the occurrence of oxidized ore in the Hypotheek mine at depths more than 700 feet below the present water table should be explained as due to valley filling of at least this amount, though in an area of great postmineral faulting such an explanation of relations occurring in only one of many lodes can not be advanced with absolute assurance.

It is noteworthy, as pointed out by Ransome,[85] that "oxidation has attacked first those parts of the ore containing pyrite and sphalerite, the solid masses of galena proving more resistant to atmospheric agencies." In partly oxidized ore cores of galena are common where no residual masses of sphalerite and pyrite remain.

---

[84] Op. cit., p. 133.
[85] Idem.

GENESIS OF THE DEPOSITS.

PERIODS OF MINERALIZATION.

The discussion of ore genesis by Ransome, with few modifications, might be incorporated as the conclusions of this report. Much confirmatory evidence, however, has been gathered during the senior writer's seven years of field work in the mining districts of Idaho. In the State as a whole two distinct periods of mineralization are recognized—one closely following the intrusion of the great Idaho batholith that crops out over a continuous area of more than 20,000 square miles in the central part of the State and the other subsequent to the extrusion of andesitic lavas that characteristically occupy deep valleys cut in an old erosion surface of postgranite age. The lavas have no important representatives north of Salmon River, but south of the river both types of deposits are well developed. The two groups are so distinct in character that even from a hand specimen the age of the deposit yielding it may generally be determined with absolute assurance. These differences are doubtless to be accounted for by a great difference in the depth at which the two groups were formed, because great difference in age, with erosion intervening by which the area was reduced to gentle topographic forms and raised several thousand feet and broad valleys as much as a mile deep were cut out and in many places flooded by lavas, can mean only that the older deposits were formed at vastly greater depth than the younger ones. It is probably safe also to assume that the older ones were deposited at much higher temperatures, because of their occurrence in the vicinity of deep-seated magmas that in several places gave off lamprophyre dikes after the ores had formed. The younger veins are characteristically fissure fillings, made up of fine-grained quartz, usually banded parallel to the walls and accompanied by small amounts of chalcedony and opal. Calcite is locally abundant and may occur either as equidimensional rhombohedral crystals or in lamellar form, the latter in most part replaced by quartz. Adularia is usually present as an accessory constituent of the gangue, and analysis of the ore usually shows selenium. Pyrite is common, but the base metals are rarely present. Examples of such deposits in the State are the veins of the Silver City, De Lamar, Singiser, Rabbitfoot, Era, and Yankee Fork districts. In each of these districts the veins are inclosed in lava rocks and are valuable for gold and silver.

In contrast to these veins the older deposits are characterized by much replacement of the wall rock, a lack of crustification, a variety of sulphides any one of which may be present in commercial amounts, and a gangue in which either siderite or massive quartz may predominate. The deposits in several districts, notably the Coeur d'Alene, Seven Devils, Mackay, Wood River, Spring Mountain, and South Mountain, have contact-metamorphic phases. Very few of them are typical fissure fillings, although fissures and shear zones for the most part directed the ore-depositing solutions so that the deposits are characteristically tabular in outline. They are valuable for lead, zinc, gold, silver, copper, antimony, and tungsten, and sometime they will be worked also for iron, nickel, and cobalt. The deposits are inclosed in granite and older sedimentary rocks.

From the above brief summary it appears that all the deposits of Shoshone County belong to the older of the two periods of mineralization recognized in Idaho.

ORIGIN OF THE FISSURES.

The structural features of the area may be grouped broadly in order of age as folding along northeast-southwest axes, faulting along north-south axes, and later faulting and folding along west-northwest axes. The faults of the third group are in part older than the ore deposits, in part younger, and in part of the same age. Normal and reverse faults occurred in both the second and third periods of diastrophic activity. The normal faults in general have dips between 50° and 80°, but the Dobson Pass fault, with a minimum displacement of 8,000 feet, has a dip of only 30°. Concerning the reverse faults Calkins[86] states:

A part only, as the Ucelly Gulch and Carpenter Gulch faults, have the low dips generally considered characteristic of overthrusts; but a far greater number clearly seen in mine workings have dips of 50° to 80° and although having displacements amounting to thousands of feet are accompanied by so little crushing and disturbance that it is impossible to conceive of their having been formed by lateral pressure alone.

Hershey[87] recognizes this fact in the Wardner area and observes that "the normal faulting made stronger gouges than reverse faulting." This observation, made by both Calkins and Hershey and fully borne out by the writers' field work, is most important, because it indicates that vertical rather than horizontal stresses operated in developing the reverse faults. These faults were formed by a lifting force which lessened the pressure of the hanging wall against the footwall instead of increasing it, as would a horizontal thrust of sufficient force to cause an upward gliding. This vertical thrust may reasonably be credited to the underlying batholith which Ransome postulated as a source of the ore-forming solutions.

[86] Ransome, F. L., and Calkins, F. C., The geology and ore deposits of the Coeur d'Alene district, Idaho: U. S. Geol. Survey Prof. Paper 62, p. 62, 1908.
[87] Hershey, O. H., Genesis of the lead-silver ores in Wardner district, Idaho: Min. and Sci. Press, vol. 104, p. 30, 1912.

The west-northwest folding is conspicuous only in the vicinity of the Osburn fault and south of it. Two anticlines cross Big Creek valley, and an overturned fold parallels the Osburn fault from Big Creek to Wallace. This is overturned northward, implying a thrust from the south. South and east from Wallace a broad syncline, with similar strike, crosses the north fork of St. Joe River and Slate Creek. It seems significant that all these folds are in the hanging wall of the Osburn fault (pp. 11–13) and that they are most marked opposite a pronounced southward bend in the fault plane. When it is recalled that the Osburn fault plane is broadly sinuous in strike, the question arises how far the disturbed zone which in general parallels it may have developed incident to the 15 miles or so of movement along the curved plane. The walls must have been in places crowded forward into reentrants and elsewhere crowded back by protuberant segments. In the Wardner district Hershey[88] has recognized an earlier and a later movement on the Osburn fault separated by the development of the New Era fault. Where cut off by the Osburn fault in the Senator Stewart mine the plane of the New Era fault is bent from a westward to an eastward dip and its strike turns to the northeast, showing drag phenomena on a tremendous scale and indicating a westward shifting south of the Osburn fault. The Wardner area comprises a great number of thin slices, which stand on edge and converge with the Osburn fault toward the west. If these structural features are incident to movement along the major fault it would seem that they imply shifting of each slice toward the northwest relative to its neighbor on the northeast (fig. 1, p. 13). If the Wardner area is moved to a position near Wallace it will be seen that protuberant parts of the Osburn fault plane are brought opposite. Furthermore it is probable that in readjustments along a curved plane by far the greater disturbance will take place in the more weakly buttressed hanging-wall side. Thus in moving westward the Wardner area was forced to conform to a relatively fixed plane which curved northward away from it. Such a condition is favorable to northwestward thrusting in the hanging-wall block. The sliced condition of the area favors such an explanation, but it is even more strongly favored by the absence of similar slicing in adjoining areas whose detailed structure is known and where mining operations afford comparable records. To the alternative conception that the intense faulting of the Wardner area is due directly to regional stresses effective in different directions at different times, such that nine distinct systems of faults may be recognized, three of them reverse and the rest normal, as Hershey believes, the localization of the phenomena is a serious objection. If the Wardner

[88] Hershey, O. H., Origin and distribution of ore in the Coeur d'Alene, p. 4, 1916.

faulting, in major part, was conditioned by movement on the Osburn plane, it may be considered in every way comparable, though on a grand scale, to the shearing and adjustment of blocks in a coarse fault gouge. Under the conditions of this conception the significance of the several systems of faults is not great, and all may be considered as incident to one general application of stress effective in movement at intervals separated by periods of accumulation of stress.

Mineralization persisted throughout much of this general period of faulting, but ore deposition principally accompanied the faults of reverse displacements. These have been shown to have less gouge than the faults of normal type, and this is sufficient reason for their being favorable channels for ore-depositing solutions. The reason for less gouge along them would seem to be the operation of an upward thrust on the hanging-wall side or a subsidence of the footwall block. As much of the faulting took place during the general period of magmatic activity, as shown by the relation of the faults to dikes, it is perhaps most reasonable to suppose that readjustments in the magma caused the vertical stresses implied by the peculiar kind of reverse faults here found. Thus there is a basis for thinking that ore deposition, which is younger than the granite and older than its lamprophyre derivatives, accompanied stages of particular activity within a magma underlying the region. This view is much different from that of Hershey,[89] who believes " that the compressive stresses necessary to make great thrusts forced the waters universally present in the rocks at great depth to ascend to the surface." It would seem that the absence of pronounced gouges on the reverse faults can only mean an absence of great compressive stresses during their development.

In the Burke area and in the vicinity of Mullan it seems that stresses causing a nearly vertical slicing of the rocks were operative. In many of the lodes in these parts of the area it is impossible to distinguish sharply between fissuring and shearing. Cleavage parallel to the lodes occurs throughout the long crosscuts leading to the Corning and Gold Hunter lodes, and schistosity parallel to the lode is conspicuously developed in the Hecla mine. In the region about Murray a similar phenomenon may be recognized, though here it is much less pronounced. It may be observed as a generalization that the further removed the deposits are from the Osburn fault the less pronounced is shearing and the more prevalent is simple fissuring.

The movements that caused the shearing were evidently recurrent, as shown by the distribution of later ore minerals in seams traversing older minerals and by the way in which the minette

[89] Hershey, O. H., Genesis of lead-silver ores in Wardner district, Idaho: Min. and Sci. Press, vol. 104, p. 827, 1912.

dike in the Hecla mine splits the ore body lengthwise. A number of the lodes also are offset by faults showing postmineral movements of considerable magnitude that apparently developed entirely new planes of dislocation.

SOURCE OF THE ORES.

The remarkable persistence of the Coeur d'Alene ore bodies in depth (Pl. XVI), their great size, the absolute gradation between the several types, possibly excepting the gold veins, the same sequence of mineral deposition in all of them, their gradation into contact-metamorphic deposits, and their localization in certain parts of the county must all be taken into account in an attempt to explain their origin.

From the wealth of observed facts about the occurrence of ore deposits that has been reported during the last half century and particularly during the last 20 years it is reasonable to assume that the deposits of Shoshone County were formed by hot ascending solutions and that the metals in them came either from an underlying magma or from material minutely disseminated in the sedimentary rocks. The evidence bearing upon these alternatives will be reviewed briefly.

The principal mines of the area comprise two groups—one north of the Osburn fault and another about 15 miles to the west and south of it. This fault, from several lines of evidence (p. 12), is believed to have a horizontal component of displacement of perhaps 15 miles to the west on the south side. If the Wardner area is set east to its former position near Wallace it is seen that the deposits are grouped with respect to a broad anticline that crosses the region in a northeasterly direction. This structure is shown by a central belt of Prichard rocks, greatly interrupted by cross faulting, which is bordered on the northwest and southeast by successively younger formations. A narrow belt along the axis of the anticline embraces the several areas of granitic rock north of Wallace. Farther south in the county similar masses of grantitic rock break through the Newland formation. Between the Osburn and Placer Creek faults no granite is exposed, although its presence at no great depth is suspected by contact phenomena in the vicinity of Wardner. The general anticlinal structure is definitely shown between these faults.

The principal bodies of ore occur on the eastern flank of this anticline, in the general zone represented by the upper part of the Prichard, the Burke, and the Revett formations but locally extending into the Newland formation. Broadly also the lead deposits lie east of the zinc deposits. Pine Creek is nearer the center of the fold than Wardner, and Ninemile Creek is nearer than Burke.

Within a few miles north and south of the Osburn fault the ore deposits are in rocks sliced by innumerable minor faults (fig. 3, p. 52), only a few of which have sufficient throw to be recorded at the surface. The ore bodies follow the minor displacements rather than the major ones, and those with reverse throw are particularly favorable. It is clear from these structural relations that the highly mineralized part of the county embraces the area about the intersection of a zone of elevation, accompanied by granitic intrusion, and a zone of exceptionally strong faulting. If the ores are of igneous derivation this accounts for their localization, but if they are of sedimentary origin it would seem that other parts of the county, particularly fractured areas of synclinal structure, might be more favorable.

The gradation between the several types of ore has been discussed elsewhere (pp. 130–132), and the reasons have been stated for concluding that all the ores of the county were formed during one period of mineralization (pp. 132–134). Subsequent discussions therefore may start with the conclusion that only one period of mineralization is represented and that the several types of deposits are merely different expressions of a common phenomenon due to a common cause.

At several places in the county (see p. 18) ore occurs in granitic rock, and in a few places lamprophyre dikes cut across the ore. As the lamprophyre dikes must be regarded as basic differentiates given off by deeper, molten portions of the intrusive mass after its outer part had solidified, it may be concluded with reasonable assurance that ore deposition took place during the period of magmatic activity in the region. This view is supported by gradation of the deposits into contact phases, as in the Success, Rex, and Helena-Frisco mines. It is also in keeping with the relation of the ore deposition to reverse faults in the Wardner district—faults along which the absence of thick gouge can be explained only by vertical stresses, and these are most likely due to an underlying mobile magma.

THE SUCCESS DEPOSIT.

A fundamental requisite to a correct conception of the genesis of the ores of the region is an understanding of the Success deposit (formerly the Granite). The Success mine is in the central part of a narrow tongue of Prichard slate and quartzite which extends eastward nearly a mile into the monzonite mass west of Gem. In the vicinity of the mine the sedimentary beds are metamorphosed to biotite schist, micaceous quartzite, and dense greenish quartzite, each of which contains variable amounts of garnet, pyroxene, biotite, muscovite, and locally chlorite and epidote. The structure of the beds is in many places difficult to decipher, but in general they strike northwest and dip 30°–85° SW. The Prichard rocks are intricately traversed on nearly all the 13 levels of the mine by tongues and dikes of the monzonite, as is well illustrated in the 300-foot level (fig. 2,

p. 26). In places the monzonite apophyses are traversed by the ore, as is clearly shown between stations 300 and 308 on this level, where a 10-foot dike nearly at right angles to the course of the ore body is exposed in opposite sides of the stope at a point where ore was extracted to a width of about 12 feet. Another equally clear example of ore formed subsequently to the monzonite appears on the same level near station 324, where an offshoot from the main ore body replaces a 4-foot dike and extends into the quartzite beyond.

On the 100-foot level, 191 feet above the 300-foot, the ore body, retaining its width of $3\frac{1}{2}$ to 5 feet, crosses a southwestward-dipping monzonite dike. This relation continues for 60 feet below the level, but lower down no ore has been found in the monzonite, although ore has been mined in the adjacent quartzite for more than 100 feet farther down. Above the level the stope continues between granite walls for 40 feet without diminution in width. As there is notably less fracturing within the dike than beyond it, the relations are interpreted as indicating this sequence—(1) shearing, (2) intrusion of the dike, and (3) renewed shearing, accompanied or followed by replacement of wall rock, whether quartzite or monzonite by ore.

The relations below the 100-foot level described above suggest that the quartzite was more easily replaced by the mineralizing solutions than the monzonite, which is clearly shown by hand specimens to have been locally replaced. This suggestion is borne out by the deposit as a whole, for, although the monzonite has been intensely altered in places, the total volume of it which has been replaced by ore, compared with the other rocks, is negligible. Of the several types of rock in the mine, the mica schist contains most of the ore, the micaceous quartzite a notable amount, the dense greenish quartzite very little, and the monzonite still less.

Unlike contact zones in limestone, the deposit here shows no sharp line of demarkation between rock that is clearly of metamorphic origin and material that is unmetamorphosed. All the invaded rocks are aphanitic. Where metamorphism is intense the texture is little changed, and difference in color, combined with a slightly increased specific gravity, is the principal suggestion that transformations have taken place. Where the principal metamorphic mineral is garnet, the color is either pink or light amber; where it is pyroxene or epidote the color is greenish. Minor amounts of these minerals, however, do not affect the normal gray or greenish-gray color. The product of intense metamorphism of the monzonite is a rock somewhat resembling the intensely metamorphosed quartzite but in most places retaining sufficient of the monzonite pattern to be easy of identification. It also has a dull appearance, due to advanced alteration in feldspar areas. Thus within the deposit two common types of alteration are recognized—exomorphism, or the transformation of the invaded rock, and endomorphism, or the metamorphism of the intrusive rock. These phenomena are described separately on pages 38–41.

The suggestion has been made that more than one period of magma injection is represented by the granitic rocks of the region, and that the metamorphism accompanied the nonhornblendic type, which is the older.[90] The specimen illustrated in Plate VI represents the hornblendic variety from the Success mine. The Lily specimen represents the syenite or nonhornblendic variety. It is clear, therefore, that the sulphides were deposited after the intrusion of both varieties of granitic rock. For this reason any difference in age of the two does not affect the problem of ore genesis. It may be doubted, however, whether more than one period of granitic intrusion is represented by the rocks of the region. After a summary of the evidence Calkins[91] states:

The conclusion drawn '* * * is that the porphyritic syenite formed as a somewhat heterogeneous marginal segregation along the southern edge of the principal plutonic mass and that it was broken up and intruded by the monzonite, perhaps before being solidified.

During the writers' field season of 1916 nothing was seen that opposes the conception of one general period of magma injection.

The evidence discussed above leads to the conclusion that the Success deposit was formed after at least the marginal solidification of the monzonite; that it is a normal type of contact deposit except for minor differences due to the quartzitic composition of the metamorphosed rock and the distribution of a thin bed which has proved more susceptible of replacement than the other formation; and that, for reasons which follow, it should be compared with such deposits as Velardeña,[92] Dolores,[93] Silverbell,[94] Mackay,[95] and Whitehorse[96] in point of genesis. In each of these districts the lime silicates and sulphides are intimately intergrown and in part replace the igneous rock. The ore deposits of each of them have been considered due to differentiates from the partly solidified intrusive. In the Mackay deposit particularly favorable relations make it susceptible of rig-

[90] Hershey, O. H., op. cit., p. 19.
[91] Op. cit., p. 46.
[92] Spurr, J. E., and Garrey, G. H., Ore deposits of the Velardeña district, Mexico: Econ. Geology, vol. 3, pp. 688–725, 1908.
[93] Spurr, J. E., Garrey, G. H., and Fenner, C. N., Study of a contact-metamorphic ore deposit—the Dolores mine, at Matehuala, S. L. P., Mexico: Econ. Geology, vol. 7, pp. 444–484, 1912.
[94] Stewart, C. A., The geology and ore deposits of the Silverbell mining district, Ariz.: Am. Inst. Min. Eng. Trans., vol. 43, pp. 240–290, 1912.
[95] Umpleby, J. B., The genesis of the Mackay copper deposits, Idaho: Econ. Geology, vol. 9, pp. 307–358, 1914.
[96] McConnell, R. G., The Whitehorse copper belt, Yukon Territory: Canada Geol. Survey Pub. 1050, 1909.

orous proof that the magma supplied most of the material of the ore and associated contact silicates, and in each of the others the evidence points strongly in the same direction. Considered as a group and in the light of their collective evidence, there can be little room to doubt that each of the deposits is due to magmatic emanations which escaped, in large part at least, after the marginal solidification of the magma. The Success deposit resembles these others in the kind of metamorphic silicates developed and their relations to the sulphides; in the age relationship of the ore minerals and the solidification of the adjacent igneous rock, and in the distribution of ore bodies, except as modified by a thin bed of exceptional shearing that was particularly susceptible of replacement.

Few deposits considered alone afford conclusive proof of their genesis, and in this particular the Success deposit is not unique. It has been shown that in all essential points it accords with the criteria of contact-metamorphic replacement deposits, and that in the sequence of events which have occurred in its development it agrees with certain well-known members of that class. The alternative view that it is a replacement phenomenon later in age than the monzonite and related to it only remotely, if at all, may be entertained by some students of the subject but is strongly opposed by what is known of the genesis and geologic occurrence of the minerals characteristic of the deposit. That the ore is not older than the monzonite is conclusively shown, in the writers' judgment, by field relations and by polished surfaces and thin sections of the monzonite described above. Hershey, however, considers the argument inconclusive and presents the alternative view in an able discussion [97] of a preliminary paper by the senior writer.[98]

In conclusion the marked concentration of mineralization near the intersection of an elevated axis of igneous intrusion and a zone of tremendous faulting, the gradation of the several types of lodes, the local development of contact-metamorphic deposits, and the definite sequence of mineral deposition in all the deposits of the county lead the writers to conclude that the ores are of magmatic derivation and that their source lies in the granitic intrusive which is exposed locally in the region and which in depth is doubtless continuous with the great Idaho batholith. This conclusion finds strong support in the fact that the relation of the deposits to the granitic rock and the late differentiates from the magma fixes the ore deposition in the general period of magmatic activity.

[97] Hershey, O. H., Genesis of Success zinc-lead deposit: Econ. Geology, vol. 12, pp. 548–558, 1917.

[98] Umpleby, J. B., Genesis of Success zinc-lead deposit: Econ. Geology, vol. 12, pp. 138–153, 1917.

The explanation also finds support in the occurrence of the Wood River deposits described by Lindgren [97] and the senior writer [100] and in the conclusive evidence in the Mackay deposits.[101] Each of these deposits is genetically related to an outlier of the main Idaho batholith.

FUTURE OF THE COEUR D'ALENE DISTRICT.

The future of the Coeur d'Alene district is perhaps as bright now as it was when Ransome examined it in 1904. A few of the large mines, such as the Tiger-Poorman and Standard-Mammoth, that were productive then have since been exhausted, but the output from others has increased so that the production of lead, silver, and zinc from the district has risen steadily. In the future zinc will cont:nue to increase, as the many zinc mines recently opened give promise of yielding a tonnage of ore comparable to that from the lead mines that have made the district famous. It does not seem likely that the lead-silver output will greatly increase from known deposits unless special economic conditions encourage more rapid exploitation during short periods. New bodies of ore, however, are being found almost every year, so that the turning point in the district's lead production is probably several years in the future. The Bunker Hill, Hercules, Hecla, and Morning mines seem to have a long period of production ahead of them, and the Tamarack and Custer is just being opened on a large scale.

The future of the copper industry in the county, as was predicted by Ransome,[102] is not so bright, but several outlying districts may develop mines of some importance, though not of a size likely to duplicate the output of the Snowstorm mine during its period of activity.

The developments of the last decade and a half have shown that the nature of the inclosing rock is not as important as appeared from the geologic distribution of developed deposits at the time of Ransome's visit. At that time it seemed that ore bodies of large size could not be expected in the Prichard formation, but since then the Interstate-Callahan and several mines in the Pine Creek area have revealed large bodies of zinc ore in this formation. So far lead deposits of note have not been found in Prichard rocks, but the writers see no valid reason why they should not occur and believe that pros-

---

[99] Lindgren, Waldemar, The gold and silver veins of Silver City, De Lamar, and other mining districts in Idaho: U. S. Geol. Survey Twentieth Ann. Rept., pt. 3, pp. 190–231, 1900.

[100] Umpleby, J. B., Ore deposits of the Hailey quadrangle, Idaho: U. S. Geol. Survey Bull. (in preparation).

[101] Umpleby, J. B., Geology and ore deposits of the Mackay region, Idaho: U. S. Geol. Survey Prof. Paper 97, 1917.

[102] Op. cit., p. 140.

pecting in Prichard areas should not be discouraged. It seems to them that any promising prospect is worthy of exploitation, regardless of the formation that incloses it. The greatest argument against prospecting in the Prichard formation is found in the mines near Burke, which give out near its upper contact. It may be wondered, however, if the control here is not structural rather than lithologic. It is very noticeable in the Standard Mammoth lode, for example, that the formations are much tighter on the lower levels than on the upper ones. The lodes here occur high on the side of an anticlinal fold, so that with increasing depth zones of less tension are being penetrated; carried sufficiently deep the tensional stresses in the outer rocks may pass into compressional stress within the anticline.

INDEX.

○

www.ingramcontent.com/pod-product-compliance
Lightning Source LLC
Chambersburg PA
CBHW032000190326
41520CB00007B/306